RAND NATIONAL DEFENSE RESEARCH INSTITUTE

Designing Unmanned Systems with Greater Autonomy

Using a Federated, Partially Open Systems Architecture Approach

Daniel Gonzales, Sarah Harting

T0308382

Prepared for the Office of the Secretary of Defense

For more information on this publication, visit www.rand.org/t/RR626

Library of Congress Cataloging-in-Publication Data is available for this publication.

ISBN: 978-0-8330-8606-8

Published by the RAND Corporation, Santa Monica, Calif.

© Copyright 2014 RAND Corporation

RAND® is a registered trademark.

Cover photo: A British MQ-9 Reaper (U.S. Air Force photo by Senior Airman David Carbajal/Released).

Support RAND
Make a tax-deductible charitable contribution at
www.rand.org/giving/contribute

www.rand.org

Preface

The U.S. Department of Defense has made substantial progress in the deployment of more capable sensors, unmanned aircraft systems (UAS), and other unmanned systems (UxS). Innovative UxS can affect the way important missions will be conducted yet impose a number of interoperability and integration challenges that must be addressed before their employment in joint operations can be effective. In addition, to provide effective capabilities in more demanding missions and environments, UxS will require more autonomous capabilities than exist in current unmanned systems. This report focuses on the architectures that define such systems and proposes a way forward built on existing efforts to improve UAS and UxS interoperability and autonomy across the joint community from the bottom up.

This research was sponsored by the Office of the Under Secretary of Defense for Acquisition, Technology, and Logistics (OUSD AT&L) and conducted within the Acquisition and Technology Policy Center of the RAND National Defense Research Institute, a federally funded research and development center sponsored by the Office of the Secretary of Defense, the Joint Staff, the Unified Combatant Commands, the Navy, the Marine Corps, the defense agencies, and the defense Intelligence Community.

For more information on the RAND Acquisition and Technology Policy Center, see http://www.rand.org/nsrd/ndri/centers/atp.html or contact the director (contact information is provided on the web page).

Contents

Figures and Tables

Figures

Tables

Summary

In 2008, U.S. Department of Defense (DoD) Secretary Robert Gates pushed the military services to field more intelligence, surveillance, and reconnaissance (ISR) assets in an effort to address the warfighter's insatiable appetite for the information these systems provide. Since that time, the DoD has made substantial progress in fielding more, and more capable unmanned aircraft systems (UAS) to meet the needs of warfighters in different theaters of operation.[1] Innovative UAS platforms and sensors have been introduced by the defense community in the last decade to meet urgent operational needs to include systems with greater endurance and improved sensors. However, innovation has also led to development of multiple programs with different communications systems, which can contribute to interoperability problems and limit both the degree to which information collected by these systems can be shared and how these systems can work together, with other systems, and be controlled by warfighters in different units or military services.

Today, unmanned systems encompass more than unmanned aircraft to include unmanned vehicles (UxVs) and unmanned systems (UxS) that operate on land and at sea. Such systems are having an effect on the way important missions are conducted, yet they also introduce a number of challenges: the number of personnel needed to operate and manage the increasing number and type of unmanned systems, the survivability of these systems against new threats, and the ability of these systems to operate in more complex and contested environments.

These challenges can potentially be met by fielding unmanned systems with greater autonomy, but to date, as noted by the Defense Science Board (DSB), such progress has been limited.[2] There are also tradeoffs associated with prioritizing the rapid fielding of advanced technologies in the near term over efforts to ensure that these new technologies are fully integrated and interoperable over the long term with existing command and control (C2) structures, weapons systems, and architectures,

---

[1] UAS includes both unmanned aerial vehicles (UAVs) and the ground control station (GCS) used to control the UAVs. We use the term UxS to include UAS, unmanned ground systems (UGS), and unmanned maritime systems (UMS).

[2] Defense Science Board, "Task Force Report: The Role of Autonomy in DoD Systems," Washington, D.C.: Office of the Secretary of Defense for Acquisition, Technology, and Logistics, July 2012.

or even just with each other. In particular, the integration and interoperability of UxV C2 systems and data links pose more complex challenges as more systems are developed and fielded. Furthermore, the growing number of UxVs increases the load and pressures on associated communications networks. One way to reduce UxV communications demands is to develop UxVs with greater autonomy. UxVs can potentially be equipped with different types of autonomous functions to reduce messaging loads on communications links to C2 and information analysis centers. For example, autonomous onboard planning algorithms can help reduce communications loads and lessen the need for frequent maneuver, heading, or flight commands. Autonomous sensor data processing algorithms can perform sensor fusion functions such as automated geo-registration and multisensor processing, can select small image chips from full images to communicate, and can even potentially automatically detect and track targets using pattern recognition or other techniques to reduce the need for communications bandwidth and human image and video analysts.

Unmanned systems will also require improved interoperability and greater autonomy to be able to operate effectively in more demanding environments and to be used more flexibly over a wider range of complex missions. However, as noted above, the DSB has found that DoD acquisition programs developing unmanned systems have made only limited progress in inserting greater autonomy into their products.[3]

Study Objectives and Scope

The objectives of this study are to identify key features of unmanned systems and their associated architectures that can enable

- improved UxV-to-UxV interoperability
- greater autonomy in unmanned systems
- cooperative UxV behaviors so that UxVs can work together in teams and accomplish complex missions in demanding anti-access/area denial (A2/AD) environments.

These objectives can potentially be accomplished by the adoption of open system architecture principles.

Initially, the scope of this research was set broadly to cover all types of unmanned systems: UAS, UGS, and UMS and joint or common architecture development efforts associated with each type of system. However, DoD research and development (R&D) efforts and joint architecture developments were found to be less mature for UGS and UMS. In addition, the navigation problems of unmanned ground vehicles (UGVs) and

---

[3] Defense Science Board, 2012.

unmanned maritime vehicles (UMVs) differ significantly from those of UAVs. These can and should influence how architectures are defined for these systems. Many important aspects of the software architectures of UGS and UMS were found to be proprietary. For these reasons, the research team focused its more detailed analysis efforts on UAS and UAS architectures. We believe that further useful work can be done on UGS and UMS architectures that can improve their interoperability and autonomous capabilities in follow-on studies.

Although significant effort has been expended to improve unmanned system interoperability, progress has been slow. In this report, we review some of the DoD's initiatives for improving UxV interoperability and autonomy and other commercial initiatives for developing autonomous vehicles. The majority of UAS program of record (POR) architecture products are currently limited to single program views, which are not easily used to assess interoperability issues that extend across program or military service boundaries. In this study, we examine how UxV architectures can be improved so that they can support efforts to increase the degree of autonomy and interoperability of unmanned systems.

Challenges to UxS Architecture Development

System architectures are complex, and ensuring compatibility and interoperability among them has been a long-standing challenge for the DoD.[4] In the past, the DoD has pursued efforts to develop a single common integrated joint architecture for a few selected mission areas. However, such efforts have proven to be too costly and difficult to complete. These difficulties resulted from technical and administrative challenges associated with architecture development and also because it was too costly for acquisition programs to change or develop new architecture products. Consequently, in the past, acquisition programs shied away from participating in joint common architecture developments.

Complete and accurate joint architectures should be developed collaboratively by all relevant military services and with the participation of at least the key acquisition programs in a joint mission area. Such participation and collaboration were difficult to achieve using the architecture approaches and development tools available in the past.

UxS programs have been no different this regard. Each UxS program has developed its own custom architectures, so any effort to develop a joint common architecture using traditional approaches would be additive and costly to existing efforts. Furthermore, to use architectures in the UxS design process, they would have to be shared

---

[4] Architectures are defined as "The structure of components, their relationships, and the principles and guidelines governing their design and evolution over time." See Department of Defense Instruction (DoDI) 4630.8, *Procedures for Interoperability and Supportability of Information Technology (IT) and National Security Systems (NSS),* enclosure 2, Washington, D.C.: Department of Defense, June 30, 2004.

electronically (similar to databases) and although architecture data are electronic, the programs often use different and noninteroperable applications. These electronic information-sharing limitations were key shortcomings of version 1.0 of the Department of Defense Architecture Framework (DoDAF).

DoDAF 2.0 is focused on developing common architecture standards to enable the electronic sharing and processing of architecture products; however, some of the tools needed for DoDAF 2.0 are not yet mature, and the DoD is understandably hesitant to endorse a single commercial application for developing architectures. Nevertheless, DoDAF 2.0 is a good and necessary first step in the process for moving DoD architecture products to interoperable extensible markup language (XML)–compatible formats.[5]

To fully exploit the capabilities of DoDAF 2.0 and given the challenges and past failed efforts to develop common joint architectures, some in the DoD have since proposed the adoption of a federated architecture approach. As described in the body of this report, such a development approach has been successfully adopted in commercial information-technology markets and has enabled innovation and collaboration among a wide range of large technology firms and small businesses.

The DoD defines a federated architecture as:

> A loosely coupled collection of information assets that accommodates the uniqueness and specific purpose of disparate architectures which allows for their autonomy and local governance while enabling the enterprise to benefit from their content. It provides an approach for aligning, locating, and linking disparate architectures and architecture information via information exchange standards to deliver a seamless outward appearance to users. Its content describes mission capabilities and the IT [information technology] capabilities necessary to respond to changing mission needs.[6]

The Joint Staff is dedicating attention and resources toward developing federated architectures and, in particular, the Joint Staff J6 is leading an effort called the Warfighting Mission Area Architecture Federation and Integration Project (AFIP). This effort supports the DoD's enterprise architecture vision to ensure that architecture data can be readily leveraged and used by decisionmakers, as the associated portal provides a repository for architecture data (provided by authoritative sources) and related data. It also aims to provide the joint force with an operational context to support interoperability and integration across capability areas and systems. Furthermore, AFIP provides an operational context and core reference architectures—

---

[5]   See Chief Information Officer, U.S. Department of Defense, "The DoDAF Architecture Framework Version 2.02," August 2010.

[6]   Department of Defense, "Department of Defense Global Information Grid Architecture Federation Strategy," Version 1.2, August 7, 2007.

leveraging Joint Mission Threads (JMTs)—that are "discoverable, available, traceable and reusable" across the joint community.[7] The effort also promotes best practices to guide architecture developers as they build their associated products and data sets.

A Common UxS Architecture Is Too Difficult to Develop Today

Should the DoD develop a single common architecture for all UxS? Or should it direct that all UxS architectures developed by individual UxS programs of record be federated? Or should separate joint common architectures be developed just for UAS, UGS, and UMS? Or should system architectures be federated separately for unmanned systems that operate in each domain? We found many significant differences in the designs, capabilities, and functions of UxS that operate in different domains. Therefore, we do not recommend that either a common or federated architecture be developed for all of them. Further, we found that UMS and UGS system developments are not as mature as UAS developments. Consequently, we recommend that further research be conducted on the software designs of UGS and UMS before a decision is made on whether architectures for these classes of systems be federated or before technical reference models be developed for them.

Federated UAS Architectures

The DoD has been actively investigating how UAS architectures can be improved. In this study, we reviewed the work of the Office of the Secretary of Defense for Acquisition, Technology, and Logistics (OSD AT&L)'s UAS Task Force Interoperability–Integrated Product Team (I-IPT), which is focused on developing a joint common UAS architecture and associated UAS interoperability profiles (USIPs). The UAS Task Force Horizontal Integration Working Group (HIWG) has developed an initial version of an "as-is" joint common UAS architecture (JCUA) that has proven useful in identifying and documenting UAS gaps.[8]

Another UAS I-IPT working group has developed the UAS control segment (UCS) architecture, which provides a common UAS GCS architecture designed to improve UAS-GCS cross-program interoperability. However, in our review of these two UAS I-IPT architectures, we found that they may not be aligned.

The UAS Task Force I-IPT was compelled to develop the JCUA and the UCS architecture because it found that current DoDAF 1.0 architecture products produced

---

[7]   William Piazza, "Joint Architecture Federation and Integration Project (JAFIP)," presentation to the Interoperability Integrated Product Team, July 18, 2012.

[8]   Chuck Frawley, "Horizontal Integration Working Group (HIWG) Update," presentation to the Interoperability Integrated Product Team, July 18, 2012.

by UAS PORs do not contain the information needed for joint interoperability analysis and to predict the occurrence of joint interoperability problems. However, it is unclear whether the JCUA represents the best approach in the long run for interoperability analysis, because it may not be adopted by UAS acquisition programs and because it may be difficult to maintain without the assistance of the major UAS acquisition PORs.

Although we do not recommend a federated architecture approach for all UxS, we do recommend that the DoD forgo developing a joint common UAS architecture and instead pursue the federation of existing UAS acquisition program architectures (to include those developed, maintained, and used by individual UAS PORs). But, can such architectures be federated (or made compatible), so that they can predict interoperability problems, without having a "central joint target" that individual service programs can use to guide system developments? We believe that a central joint target is needed and recommend that this be based on the UCS architecture and on a technical reference model (TRM) that we propose below. Thus, the UCS architecture should be one that is federated. We recommend it for federation because it relies on open standards, has a well defined modular structure and TRM, and it has already been useful in demonstrating cross-program UAS interoperability.

One challenge to federating UAS architectures is achieving agreement on and using a common syntax (or common dictionary of terms) for UAV and UAS functions, capabilities, and interfaces.

Incrementally Develop a Common UAS Architecture Syntax

A significant challenge to developing federated architectures is the lack of common semantics or syntax and data standards across architectures. In other words, architectures developed by individual programs may label the same components by different terms across applications. Even if there are electronic representations of the same architecture across services, the application may not recognize it as the same architecture. As a result, the program architectures cannot be easily combined or shared to assess interoperability or to be reused.

There are three options for addressing the syntax challenge. The first option is to create a translating service common across services. The second is to develop joint semantics for key architecture data elements. The third is a combination of both approaches.

We recommend that the DoD develop common joint semantics for key architecture data elements. Because of the complexity of this task, it should be approached in a top-down manner where the highest-priority modules or functional elements of the architecture are sequentially addressed in efforts to develop a common syntax. The key elements in a common taxonomy should define (at least at a top level) UAS capabilities, components or subsystems, commands or messages, and mission elements.

If this is done, these could then be used to express JMTs associated with these architectures and also to incorporate service POR architecture products into the Joint Architecture Federation and Integration Project (JAFIP). This is important because JAFIP will work only if a common syntax and data standards are developed; otherwise, a federation is not possible. In the short term, efforts should focus on semantics to make the sharing proposed by JAFIP work properly.

We recommend the third approach overall, given the difficulty in creating a comprehensive common syntax for all UAS and the time it would take to reach agreement among all programs and players. If a common syntax is developed for major UAS architecture data elements, then the number and complexity of translated terms can be minimized.

Additional Steps to Enable UAS Architecture Federation and Autonomy

Additional steps will likely be necessary to enable the federation of individual UAS architectures. If these architectures are federated, the result can lead to UAS with greater autonomy and interoperability. An important enabling step toward these goals was made a part of DoD program acquisition guidance two decades ago by the USD (AT&L). At that time, acquisition programs were encouraged to use a modular open system architecture (MOSA) approach and to use open standards in the design and specification of DoD systems. However, MOSA was never mandated and was adopted by only a few major DoD acquisition programs. Recently, the Office of the Deputy Assistant Secretary of Defense for Systems Engineering (ODASD-SE) updated MOSA guidance and renamed it open system architecture (OSA) guidance.[9] ODASD-SE OSA guidance emphasizes the use of open standards but does not require a modular system design.

Our review of UAS architecture and autonomous capability developments leads us to recommend that a high-level modular framework be adopted that can guide the development of federated UAS architectures. We call this framework a partially open systems architecture (POSA). The model for this concept is based on MOSA concepts. In MOSA, the system architecture is decomposed into key components or modules. Here, the interfaces connecting modules are required to be defined using open standards. We distinguish POSA from MOSA in that a POSA architecture decomposes only select parts of the system into components that are integrated by open stan-

---

[9]   ODASD-SE, "DoD Systems Engineering—Initiatives," December 2013.

dards.[10] In our proposed UAS POSA, only the components of a UAS that are critical for interoperability or autonomous capabilities are decomposed in this way.

A POSA framework will not only improve unmanned system interoperability but can also enable autonomous capabilities to be more easily integrated into unmanned systems after they are developed. If the interfaces of the POSA framework are designed properly, with open interfaces, then new software packages and new hardware can be inserted into the system as part of an upgrade cycle. POSA modularity will also enable program managers to reuse software code, choose the best solution for a particular autonomous system function from competing contractors, and enable contractor teams to collaborate in the development of new software-based autonomous capabilities, as has been done in the commercial world with the implementation of App stores for iOS and Android mobile platforms.

A considerable amount of work has been done on common architectures for UAS. Therefore, for this class of unmanned system, we can go further and recommend a modular TRM for UAS that is consistent with our POSA approach.

A UAS Technical Reference Model

Our review of software architectures for autonomous or semiautonomous unmanned systems reveals that different development teams have selected different modular schemes for their architectures. In addition, they have chosen different software foundations for their software architectures. This extends to the messaging standards and messaging approach that are used. For example, some software architectures rely on a centralized database or server for the messaging infrastructure, but others do not. Architecture frameworks that appear to be used most frequently in UAVs fall into the latter category. The advantage of a decentralized communications bus is that they generally have real-time messaging performance, which is important for maintaining real-time control of a fast-moving vehicle. An example of the decentralized approach is the Open Management Group (OMG) Data Distribution Service (DDS), which is used in many real-time tactical systems in the DoD.

Some additional important architectural constructs have been proposed in academic research projects that have developed autonomous UAVs. One such concept is that of multiple levels of control that has been used by a number of different research teams (the Johns Hopkins University Applied Physics Laboratory UAV architecture, the Massachusetts Institute of Technology Cooperative Search, Acquisition, and Track architecture, and the more recent UAV architecture proposed by MIT Lincoln Laboratory).

---

[10] This distinction and POSA are defined in detail in the body of this report.

In Figure S.1, we identify what we believe are the essential modular components of a UAV TRM that can support multiple levels of autonomy and which are consistent with the UAS POSA approach described above. This proposed TRM includes features that have been identified in the UAV high-level architectures developed and, in many cases, demonstrated by different research teams. The proposed TRM supports two levels of system control. Highly responsive vehicle control would be accomplished using the Auto Pilot Module (APM), which would be connected to flight control systems and sensors using a high-speed tactical data bus. This data bus would deliver messages with a high assurance of low or minimal time delay to enable real-time control and feedback loops. Other services or modules that require real-time performance would also be connected to the same tactical data bus. We recommend that the tactical service bus use the DDS standard, as indicated in the figure, to ensure interoperability with the UCS architecture.

The system would be equipped with the second enterprise service bus to support services that did not have real-time communications requirements. Examples of these are shown in the figure, including air domain services that would process air track information and sense and avoid (SAA) advisories from offboard C2 or air surveillance

Figure S.1
Proposed UAS Modular Technical Reference Model (TRM)

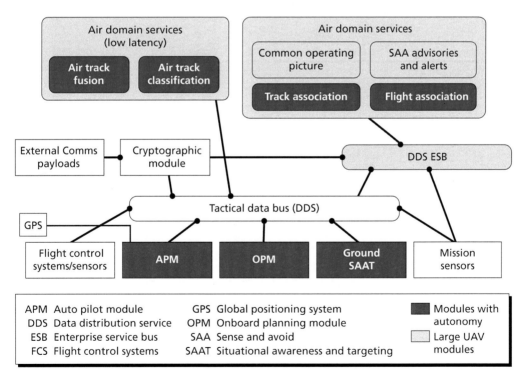

centers. In addition, mission sensors that produced a high volume of mission data also may be connected using the enterprise service bus.

The modules in the TRM that could be enabled with autonomous capability are shown in blue in the figure. These modules would be connected to the overall system using open interfaces. This would enable these modules to be produced by outside contractors that have special expertise in autonomous systems. The TRM would enable autonomy to be inserted into four key areas in the UAS architecture: (1) the real-time APM, (2) the ground situational awareness and targeting module (SAAT), (3) the onboard planning module (OPM), and (4) the low latency air track management services.

With further development and specification of the open interfaces that connect these modules to the larger system, this TRM will enable UAS programs to comply with the design recommendation made by the Defense Science Board, namely, to separate the autonomous capabilities of unmanned systems from the rest of the vehicle platform.

All of the modules highlighted in blue could reduce UAV communications demands significantly and could eliminate the need for real-time communications to the vehicle for remote pilot control in A2/AD environments. For example, air track would not have to be sent up to or down from the aircraft to identify potential aircraft or terrain collision events.

It should also be noted that the TRM shown in the figure includes all of the modules needed by a large high-value UAV that could fly in U.S. Federal Aviation Administration (FAA)–controlled airspace and that may be equipped with defensive countermeasures against aircraft threats. Smaller UAVs that fly at low altitude, that would not need to fly in FAA-controlled airspace, and that would be too small to be equipped with aircraft sensors and warning systems would not need the modules highlighted in purple in the figure.

UGV and UMV POSA Frameworks

Should POSAs and TRMs for UGVs and UMVs be similar in composition and scope to the one proposed for UAS? We believe the answer to this question is no. Unmanned systems are robots that are increasingly going to be programmed to operate semiautonomously in specific ways with warfighters and threats in distinct environments, just as ships, ground vehicles, and even traditional manned aircraft interact in fundamentally different ways.[11] We found evidence of these distinctions when examining the different architectures and modular designs of unmanned systems designed

---

[11]   We recognize that manned aircraft have an autopilot capability, which is a form of semiautonomy. However, in this case, the pilot is always in the aircraft and can assume control. In an unmanned system, pilots operate the system remotely and, furthermore, some UAS are launch-on-mission with no interaction until the UAS returns.

to operate in these different domains. Our review of high-level unmanned system architectures and frameworks reveals these frameworks to be dissimilar for unmanned systems that operate in different domains (air, maritime, and ground). The mission context, control systems for each class of system, and the environmental factors that each class of system must contend with are so different that they may lead to dissimilar architectures and frameworks. Therefore, we recommend that POSA development efforts be independently pursued for UxS that operate in different domains. We believe that a tailored POSA should be developed for UGS and a separate UMS POSA should be developed for USVs and UUVs.

Proposed Next Steps

Even though it may be premature to develop common or federated architectures for UGS and UMS, developments of common syntaxes for UGS and UMS architectures should be a near-term effort, as this will make future joint architecture developments easier regardless of which approach is eventually chosen for UGS and UMS.

Further research should be conducted on autonomous UGVs and UMVs[12] and on the architectural constructs used by developers of these systems before TRMs are developed for these classes of unmanned systems. Several important Defense Advanced Research Projects Agency (DARPA) and Navy programs are nearing development of initial conceptual designs for USVs and UMVs. Information from these programs can be used to develop a TRM for a joint common unmanned maritime system architecture that is based on the latest software and robot technologies.

In addition, Google and commercial automobile manufacturers are developing proprietary autonomous vehicle systems and technologies. The underlying architectures for these systems are proprietary and have not been made available to the open-source software development community. However, it may be possible, with assistance from major industry firms, to develop a UGV POSA that can be used to advance military UGV development, interoperability, and autonomy and that still preserves the intellectual property and R&D investments of private firms. Research in this area could be conducted by a federally funded research and development center if appropriate nondisclosure agreements are negotiated with these private firms.

In our analysis, we investigate in detail the specific software architectures used by the different development teams that have explored or developed autonomous unmanned systems. Important insights can be gained by examining and comparing the details of the software architectures. For example, should the TRM include specified software development environments and software development kits? A related

---

[12] UMVs include unmanned surface vehicles (USVs) or ships and unmanned underwater vehicles (UUVs) or submarines.

question is whether open source software code bases should be used in specific parts of an open modular architecture for UAS?

And, finally, the open interfaces should be defined for the key modules in the proposed UAS TRM that are designed to contain autonomous capabilities. These open interfaces could be established by examining the messaging formats and communication buses used by leading research teams and by interviewing autonomy experts in the DoD R&D community and industry.

Acknowledgments

This research was made possible through the support of our sponsor, William Scott, Office of the Assistant Secretary of Defense for Research and Engineering. We thank him for his leadership, guidance, and encouragement throughout the course of this project.

We also extend our thanks to the UAS Task Force in the Office of the Secretary of Defense and the UAS I-IPT team, under the leadership of CDR J. P. Greene as well as CAPT Karl Thomas before him. OSD's UAS I-IPT continues important efforts to further interoperability among unmanned systems, and our work has benefited from their insights and expertise.

We thank Cynthia Cook at RAND for her support of this research, and Randy Steeb and Sherrill Lingel, our RAND colleagues, for their careful and thoughtful reviews of the report. This report has been much improved by their efforts. Finally, we are also grateful to Lovancy Ingram for her expert assistance in the preparation of this report.

Abbreviations

A2/AD	anti-access and area denial
ADEPT	all-domain execution and planning technology
ADSB	Automatic Dependent Surveillance–Broadcast
AFIP	Architecture Federation and Integration Project
AFRL	Air Force Research Laboratory
AIXM	aeronautical information exchange model
API	application programming interface
APM	auto pilot module
ARGUS	Autonomous Real-Time Ground Ubiquitous Surveillance
ATC	Air Traffic Control
AV	autonomous vehicle
C2	command and control
CBRN	chemical, biological, radiological, and nuclear
CIT	communications and information technology
CMRE	Centre for Maritime Research and Experimentation
CORBA	Common Object Request Broker Architecture
CoT	cursor on target
COTS	commercial off the shelf
CoV	class of vehicle
CSAT	cooperative search, acquisition, and track
DARPA	Defense Advanced Research Projects Agency

DDS	Data Distribution Service
DEM	Digital Elevation Map
DoD	Department of Defense
DoDAF	Department of Defense Architecture Framework
DSB	Defense Science Board
EEI	Essential Element of Information
EOD	explosive ordnance disposal
ESB	Enterprise Service Bus
FAA	Federal Aviation Administration
GBSAA	ground-based sense and avoid
GCS	ground control station
GIG	global information grid
GOSS	government open source software
GPS	global positioning system
HIWG	Horizontal Integration Working Group
IDL	interface definition language
IED	improvised explosive device
I-IPT	Interoperability–Integrated Product Team
IOP	interoperability profile
IPT	Integrated Product Team
IS	imaging system
ISR	intelligence, surveillance, and reconnaissance
IvP	interval programming
JAFIP	Joint Architecture Federation and Integration Project
JAUS	joint autonomous unmanned system
JCUA	joint common UAS architecture
JHU APL	Johns Hopkins University Applied Physics Laboratory

JMS	Java Message Service
JMT	joint mission threat
JPO	Joint Program Office
LCS	littoral combat ship
LIDAR	laser imaging detection and ranging
MCM	mine countermeasures
MIT/LL	Massachusetts Institute of Technology Lincoln Laboratory
MOAA	maritime open autonomy architecture
MOOS	Mission Oriented Operating Suite
MOSA	modular open system architecture
MPI	Message Passing Interface
NAS	national airspace system
NCES	Network Centric Enterprise Services
NCT	Network Control Terminal
NIST	National Institute of Standards and Technology
NOTAM	notice to airmen
OBP	onboard processing
OCU	Operator Control Unit
ODASD-SE	Office of the Deputy Assistant Secretary of Defense for Systems Engineering
OGS	OpenGeoSys
OMG	open management group
ONR	Office of Naval Research
OODA	observe–orient–decide–act
OPM	onboard planning module
OS	operating system
OSA	open system architecture
OSD	Office of the Secretary of Defense

OSD (AT&L)	Office of the Secretary of Defense for Acquisition, Technology, and Logistics
OSE	Operating System Environment
OUSD (AT&L)	Office of the Under Secretary of Defense for Acquisition, Technology, and Logistics
OVM	onboard vision module
PC	physical computing
PED	processing exploitation and dissemination
POR	program of record
POSA	partially open systems architecture
R&D	research and development
RF	radio frequency
RS	robotic system
SAA	sense and avoid
SAAT	situational awareness and targeting module
SAE	Society of Automotive Engineers
SDK	software development kit
SOA	service-oriented architecture
SOF	Special Operations Force
SPS	Signalling Protocols and Switching
TA	target acquisition
TCAS	traffic collision avoidance system
TRM	technical reference model
TST	time sensitive target
UAS	unmanned aircraft system
UAV	unmanned aircraft vehicle
UCI	User-Computer Interface
UCS	UAS control segment

UGS	unmanned ground system
UGV	unmanned ground vehicle
UI	user interface
UMS	unmanned maritime system
UMV	unmanned maritime vehicle
USAL	UAV service abstraction layer
USIP	UAS interoperability profile
USV	unmanned surface vehicle
UUV	unmanned underwater vehicle
UxS	unmanned system
UxV	unmanned vehicle
WXXM	weather information exchange model
XML	extensible markup language

CHAPTER ONE

Introduction

The U.S. Department of Defense (DoD) has made substantial progress in the development and use of unmanned systems. The Army, Navy, Marine Corps, and Air Force have deployed a wide range of unmanned systems to perform a variety of missions.[1] Unmanned aircraft vehicles (UAVs) that carry weapons as well as sensors have proven especially valuable in efforts to counter terrorist groups such as al Qaeda in remote areas of the world. To be sure, armed UAVs are influencing the way such high-priority strike missions are being conducted. Despite these successes, however, the DoD faces a number of challenges in fielding unmanned systems that can operate in a wide range of environments and threat conditions and that can contribute to missions on the ground, at sea, and in the air.

The DoD has started several initiatives to improve the levels of interoperability and integration between unmanned systems. The Office of the Secretary of Defense's (OSD's) unmanned aircraft systems (UAS) Interoperability Integrated Product Team (I-IPT) is a cross-program integration effort focused on developing architectures and standards that may be leveraged by DoD programs. Yet, given that UAS technologies may be developed faster than DoD guidance and standards, are additional steps needed to ensure that DoD UAS can operate seamlessly with each other in an integrated fashion, as well as with manned platforms and commanders on the battlefield?

The Army and Marine Corps are using unmanned ground systems (UGS) and UAS in current operations and are planning to develop unmanned ground vehicles (UGVs) and UAS with greater capabilities. In addition, the Navy and Defense Advanced Research Projects Agency (DARPA) are developing advanced unmanned surface vehicles (USVs) and unmanned underwater vehicles (UUVs) for maritime environments. Although the operating environments and challenges are different for each type of unmanned system, it may be possible to glean lessons from UAS development efforts and apply them to the development of unmanned systems designed for other domains. A related question is whether the guidance and standards that have been developed for UAS can be or should be applied to the development and standardization of UGVs and

---

[1] Department of Defense, "Unmanned Systems Integrated Roadmap FY2011–2036," 2011. See also Department of Defense, "Unmanned Systems Integrated Roadmap, FY2013–2038," 2013b.

USVs, or whether these other types of unmanned vehicles require different architectural constructs to enhance or ensure their interoperability and integration.

Semiautonomous capabilities have the potential to enable mission-level command of unmanned systems and groups or swarms of unmanned vehicles to conduct complex missions in a cooperative manner independently or with limited oversight from remote human controllers.

For example, should joint UAS architecture requirements provide more specific guidance on what software application programming interfaces (APIs) or command and control (C2) messages unmanned systems should use? Should emerging joint unmanned system architectures include APIs or other types of interfaces for autonomous or semiautonomous UAS functions, including interfaces with onboard mission and flight control programs? If such information were furnished in architecture documents, it may improve unmanned system interoperability and enable them to be used more flexibly across system, unit, and joint force boundaries.

A DoD goal for the next generation of unmanned systems is to enable warfighters to seamlessly integrate them into joint warfighting missions:

> DOD envisions unmanned systems seamlessly operating with manned systems while gradually reducing the degree of human control and decision making required for the unmanned portion of the force structure.[2]

An underlying assumption related to this DoD goal is that future unmanned systems will require and possess more autonomous capabilities than current unmanned systems (which have very limited autonomy). Therefore, to enable this vision, greater forms of unmanned system autonomy are required.

Further, this vision will also require that unmanned systems be interoperable on many levels, to include the ability to dynamically share information, including situational awareness and targeting information, with other unmanned systems and with manned platforms. How should these additional levels of autonomy and interoperability be characterized? And how should they be represented in acquisition program documentation and guidance?

Should the DoD build an additional layer to the existing architecture for incorporating semiautonomous systems so that future programs can tap into a federated architecture and build toward it using prescribed standards that enable it to interoperate with other UAS and manned systems?

---

[2] Department of Defense, 2011, 2013b.

Classifications of Autonomy for Unmanned Systems

Past researchers have classified different levels of autonomy that unmanned systems could be designed to have and that ultimately could be demonstrated. These levels were designed to provide a framework for incrementally increasing unmanned system (UxS) autonomy through a stepping-stone approach and moving from one level to another. These levels of autonomy were defined by the National Institute of Standards and Technology (NIST) in conjunction with industry experts about a decade ago. NIST defines autonomy as "the condition or quality of being self-governing" and the unmanned system's "ability of sensing, perceiving, analyzing, communicating, planning, decision-making, and acting, to achieve its goals as assigned by its human operator(s) through designed HRI [human-robot interaction]."[3] The four levels of interoperability the NIST working group defined are:

- Fully autonomous: "A mode of operation of an UMS [unmanned maritime system] wherein the UMS is expected to accomplish its mission, within a defined scope, without human intervention. Note that a team of UMSs may be fully autonomous while the individual team members may not be due to the needs to coordinate during the execution of team missions."[4]
- Semiautonomous: "A mode of operation of a UMS wherein the human operator and/or the UMS plan(s) and conduct(s) a mission and requires various levels of HRI."[5]
- Teleoperation: "A mode of operation of a UMS wherein the human operator, using video feedback and/or other sensory feedback, either directly controls the actuators or assigns incremental goals, waypoints in mobility situations, on a continuous basis, from off the vehicle and via a tethered or radio linked control device. In this mode, the UMS may take limited initiative in reaching the assigned incremental goals."[6]
- Remote control: "A mode of operation of a UMS wherein the human operator, without benefit of video or other sensory feedback, directly controls the actuators of the UMS on a continuous basis, from off the vehicle and via a tethered or radio linked control device using visual line-of-sight cues. In this mode, the UMS takes no initiative and relies on continuous or nearly continuous input from the user."[7]

---

[3] Federal Agencies Ad Hoc Autonomy Levels for Unmanned Systems Working Group Participants, "Autonomy Levels for Unmanned Systems (ALFUS) Framework, Volume I: Terminology, Version 1.1," NIST Special Publication 1011, September 2004, p. 8.

[4] Federal Agencies Ad Hoc Autonomy Levels for Unmanned Systems Working Group Participants, 2004, p. 14.

[5] Federal Agencies Ad Hoc Autonomy Levels for Unmanned Systems Working Group Participants, 2004, p. 14.

[6] Federal Agencies Ad Hoc Autonomy Levels for Unmanned Systems Working Group Participants, 2004, p. 14.

[7] Federal Agencies Ad Hoc Autonomy Levels for Unmanned Systems Working Group Participants, 2004, p. 14.

The vast majority of unmanned systems employed by U.S. military forces over the past decade have very limited autonomy and thus would be classified by NIST as remote-controlled or teleoperated systems during the majority of their mission profile and for most of their mission functions. It has proven difficult to incrementally improve current unmanned systems from these lower levels of autonomy so that they have significant autonomous capabilities. Some engineering experts believe that this past experience indicates that a development framework based on levels of autonomy has serious shortcomings. To this point, the Defense Science Board (DSB) has recommended abandoning levels of autonomy and instead argues for the use of a three-facet autonomous systems framework centered on cognitive echelon, mission time lines, and human-machine system trade spaces.[8] The authors agree with the DSB that although NIST framework may be useful for classifying unmanned systems, it is not useful for guiding autonomous system development.

Study Objectives

The objectives of this study are to identify key features of unmanned systems and their associated architectures that can enable

- improved unmanned vehicle (UxV)-to-UxV interoperability
- greater autonomy in unmanned systems
- cooperative UxV behaviors so that UxVs can work together in teams and accomplish complex missions in demanding anti-access/area denial (A2/AD) environments.

As discussed below, these objectives can potentially be accomplished by the adoption of open system architecture principles. However, a challenge for the DoD and for unmanned system acquisition programs is how best to implement these architecture principles using current DoD architecture development guidance and also how to insert these principles into the joint architecture development initiatives that are now under way.

DoD Architecture Concepts

Architecture concepts for complex systems have evolved and changed considerably over time. The first modern concepts for information technology architecture can be

---

[8] Defense Science Board, "Task Force Report: The Role of Autonomy in DoD Systems," Washington, D.C.: Office of the Under Secretary of Defense for Acquisition, Technology, and Logistics, July 2012, p. 2.

attributed to John Zachman.[9] Zachman identified the need for creating architecture products for the different people or organizations that contribute to the building of a complex information system. He called these the owner, the builder, and the designer. Architecture products were to be produced for each and aligned to ensure that the expectations of all communities were met when the final product was completed.

In the context of a DoD program, these roles translate to the operator (or warfighter), the contractor, and the acquisition program office. A DoD architecture was decomposed into three types of views for these communities: first, the operational views; second, the technical views; and third, the system views. Initially, DoD architecture guidance was applied only to communications and information technology (CIT) programs. However, as the number of CIT components increased in DoD platforms, such as fighter jets and UAS, the scope of DoD architecture guidance and tools expanded to include a much broader set of systems.

DoD policy defines an architecture as:

> The structure of components, their relationships, and the principles and guidelines governing their design and evolution over time.[10]

Over time, it became apparent that the architecture products produced by acquisition programs required their own standards and a data model to enable effective architecture analysis and also allow accurate representation of the latest software technologies. Consequently, DoD architecture standards and guidance has continued to evolve over time. The current DoD Architecture Framework (DoDAF) for expressing interoperability needs and standards is DoDAF version 2.0.[11] But its adoption by programs of record has been slow because it represents a large change from earlier versions of the framework. The Joint Staff is attempting to move programs to adapt the Joint Chiefs of Staff J6 Warfighting Mission Area Architecture Federation and Integration Project (AFIP), which provides a foundation and infrastructure for improving integration and interoperability among manned and unmanned systems. However, these efforts may not be sufficient for addressing persistent interoperability challenges, and in their current form they may not enable new advanced technologies, such as autonomous or semiautonomous decisionmaking, to be incorporated in unmanned systems.

However, a comprehensive analysis of all aspects of current DoD architecture guidance and tools is beyond the scope of the current study.

---

[9] John A. Zachman, "A Framework for Information Systems Architecture," *IBM Systems Journal,* Vol. 26, No. 3, 1987, pp. 276–292.

[10] Department of Defense Instruction 4630.8, *Procedures for Interoperability and Supportability of Information Technology (IT) and National Security Systems (NSS),* enclosure 2, Washington, D.C.: Department of Defense, June 30, 2004.

[11] Architect Role and Developer Role, "The DoDAF Architecture Framework Version 2.0," 2011.

Analytical Approach

This research was motivated by several observations. First, all of the military services have invested considerable resources into the development of unmanned systems. Also, as alluded to above, the DoD has expressed the desire to improve UxS interoperability and autonomy to extend the mission space and capabilities of these systems. Nevertheless, frustrating UxS interoperability problems have been encountered in recent operations. Third, considerable research and development (R&D) on autonomous UxS has been conducted by DoD-funded contractors and academics, but with limited results. These issues served to define the approach and scope of this work.

The study team examined the interoperability and integration efforts of the DoD UAS Task Force and attended meetings of several task force working groups. We surveyed a wide range of UxS developments that were focused on developing autonomous capabilities.

We reviewed past RAND research on unmanned systems and on UAS and UGS interoperability issues. The study team also leveraged other RAND research on military operations against adversaries that may employ A2/AD tactics and the role that UAS could play in such environments.

We also reviewed past research on information technology architectures, DoD architecture frameworks and guidance, and the latest DoD architecture integration and federation initiatives.

Scope

Initially, the scope of this research was set broadly to cover all types of unmanned systems: UAS, UGS, and UMS, and joint or common architecture development efforts associated with each type of system. However, because DoD R&D efforts and joint architecture developments were found to be less mature for UGS and UMS, and because the software architectures of UGS and UMS were found to be proprietary, the research team focused its later, more detailed efforts on UAS architectures. We believe that further useful work can be done on UGS and UMS architectures that can improve their interoperability and autonomous capabilities in follow-on studies.

Caveats

This study was subject to some important limitations. The study team did not access proprietary information from UxS contractors. Consequently, we could not compare competing system designs, software, or development approaches. We also did not review actual UxS software code, software development kits (SDKs), or compilers. Much of this software code may be considered proprietary by UxS developers.

Organization of This Report

We continue this report with a discussion of the mission space for DoD unmanned systems, to include examples of how unmanned systems have been used in recent operations, potential uses in future operations, and interoperability challenges experienced to date. We then discuss ongoing OSD-led interoperability initiatives pertaining to UAS development. The fourth chapter then describes the limitations and opportunities for UAS autonomy. Next, we describe a modular open system architecture (MOSA) and related UAS architecture efforts and developments. We conclude with a summary of our analysis and a recommended way forward to support the DoD's UxS architecture federation efforts.

CHAPTER TWO

Expanding the UxS Mission Space

In this chapter, we discuss the mission space for DoD unmanned systems, first summarizing how these systems have been used in recent operations. We then discuss how they may be used in future operations and interoperability challenges.

UxS in Recent Operations

Unmanned Aerial Systems

UAS have played an important role in recent operations. UAS missions include tactical reconnaissance and surveillance and, most recently, time sensitive target (TST) strike missions in which intelligence, surveillance, and reconnaissance (ISR) and strike capabilities are integrated on the same unmanned platform. This has enabled UAVs to strike high-value targets quickly, before they move out of range or out of sight. Having both capabilities on the same platform greatly reduces the reaction time and has enabled UAS to successfully carry out TST missions.

Recent UAS operations have largely taken place in uncontested air environments where the United States has overwhelming air superiority or where adversaries do not have the capability to threaten U.S. UAS above a minimal altitude. In these operations, UAS have been under the near continuous control of U.S. operators who are sometimes within line of sight of the aircraft or in other cases in the continental United States controlling the aircraft via satellite communication links. If these communication links were to become unavailable because of adversary action, the current generation of unmanned systems would be unable to carry out the missions they are now performing in uncontested environments. Therefore, an important consideration for the DoD is how to extend the capabilities of current unmanned systems to A2/AD environments.

However, UAS are not the only type of robotic system that has proven to be valuable in military operations. Robotic systems have been developed and used by the U.S. military for other domains as well.

Unmanned Ground Systems

UGS have also provided important new capabilities in recent operations. These systems have been used to extend the reach of human operators to enable ground forces to identify and neutralize improvised explosive devices (IEDs) from a safe distance. However, UGS have played a more limited role in military operations than UAVs because of two primary challenges: first, the inability of UGS to navigate complex terrain and obstacles on the battlefield, and, second, the inability of UGS to safely employ weapons in the proximity of friendly forces without introducing the risk of fratricide. Target identification (i.e., the ability to distinguish between friendly and enemy armored vehicles), even in traditional combat environments between armored forces, has posed a challenge. Target identification in modern counterinsurgency operations is even more difficult to do in an automated fashion today with current technologies.

Unmanned Maritime Systems

UMS can operate on the ocean surface or beneath it. UMS that operate on the surface are USVs and those that operate below the surface are UUVs.

Unmanned Surface Vehicles

The use of USVs has been demonstrated in naval operations in a limited number of missions, such as countermine warfare. Remote control of unmanned vessels is more difficult from naval ships than for UAS because of the limited range of communications between Navy ships and surface vessels. To extend the range of USV operations far beyond manned vessels would require an intermediate or relay node, such as a manned aircraft, a UAS, or a satellite. Just as with current UAVs, if a way to accomplish long-range communications is not available (for example, because of a satellite failure or enemy jamming), long-range USVs would not be able to perform their mission unless they had some autonomous capabilities. All USVs developed today possess only a limited degree of autonomy, although there are research projects under way to provide a greater degree of autonomy to USVs.[1]

Despite the relatively late start in the development of USVs, the DoD is conducting significant research on USVs for a variety of missions, to include ISR, oceanographic surveillance, countermine, and small boat security missions.

Unmanned System Characteristics

Table 2.1 lists a number of unmanned platform attributes that can be used to assess the operational capabilities, missions, and C2 challenges associated with unmanned systems. From the table, one can see that the environment in which the unmanned

---

[1]   "ACTUV Program Initiates Concept Designs," DARPA press release, December 2010.

Table 2.1
Comparison of UxS Designed for Different Environmental Domains

Attribute	UAS	UGS	UMS
Endurance	USVs have advantages over UAS, particularly when operating at low speed	Relatively limited	UMS typically have the greatest endurance; USVs have advantage over UUVs
Speed	Greater speeds possible than UMS or UGS	Limited (less than 100 km/hr)	UUVs are relatively slower by a few knots than USVs; USVs speeds are lower than UAS or UGS speeds
Launch and recovery	Unique takeoff and landing risks; additional sensor data may be needed	Relatively straightforward; no different from rest of route	Port and harbor operations rules of navigation needed
Navigation	GPS or inertial guidance to determine vehicle position sufficient in most cases	Terrain and road maps needed in addition to vehicle position	GPS or inertial guidance to determine ship position and depth charts sufficient except along shipping routes and in port
Remote control response times	Milliseconds	Milliseconds	Seconds to minutes
Control surfaces	Wings, ailerons, flaps, and rudders	Tires, tracks, or feet	Fins, rudders, hatches, and propellers
Range	Greater range than UMS	Limited to typical motor vehicle ranges (~ 300 km or less)	USVs have greater range than UUVs
Payload capacity	Limited space, weight, and power even for large UAS; very limited for small UAS	Limited, especially for UGVs capable of off-road travel	USV have high payload capacity; UUVs have low energy density, reducing internal volume for payloads
Mission areas	Penetrating, persistent, tactical, small tactical, micro/mini	EOD, CBRN, protection, engineer, logistics, transport, ISR, C2	MCM, maritime security (ISR, port surveillance, SOF support, electronic warfare)
Stealth	Some potential	None	Some potential

SOURCES: Department of Defense, 2013b; Scott Savitz, Irv Blickstein, Peter Buryk, Robert W. Button, Paul DeLuca, James A. Dryden, Jason Mastbaum, Jan Osburg, Philip Padilla, Amy Potter, Carter C. Price, Lloyd Thrall, Susan K. Woodward, Roland J. Yardley, and John Yurchak, *U.S. Navy Employment Options for Unmanned Surface Vehicles (USVs)*, Santa Monica, Calif.: RAND Corporation, RR-384-NAVY, 2013, p. 27.

system operates can have a significant influence on required system capabilities and characteristics. Also, it should be noted that although manned aircraft and ships can be relatively large—for example, commercial jumbo jets used for passenger travel, super tankers for oil transport, or Navy aircraft carriers—UAS and USVs have to date been relatively small platforms with limited payload capacity, as indicated in the table. The speed of UMS and UGS is limited by the environment in which they must operate and by the shape of the hull or size of ground vehicle. Ship architects and automobile engineers have long experience in fitting efficient propulsion systems into ships and ground vehicles to maximize speed, payload size, or endurance (but not all three attributes simultaneously). In contrast, UAS can fly at much faster speeds, especially if they are equipped with jet engines, but with payloads of relatively small size. Of course, vehicle designers have the freedom to design vehicles over a relatively large trade space and can choose to maximize one or two of the above attributes at the expense of a third one (although aircraft must be able to fly at a minimum speed to avoid stalling and enable takeoff). These differences in propulsion, platform shape, and minimum cruising speeds have important implications for the control system response times (as indicated in the table).

Some other important characteristics of the system shown in Table 2.1 include how the system is launched and recovered and how the vehicle navigates. In these two areas, UAS, UGS, and UMS all present their own unique challenges, either for remote control of the vehicles or even if the systems were to operate autonomously. These challenges are due to the unique and different environments in which the vehicles must operate, especially in a peacetime or civilian traffic environment. For example, operating in commercial airspace or in busy harbors or shipping lanes will require additional capabilities specified by regulators. Such regulations are currently in development for UAS but need to be developed by regulators for UMS. We will examine some of these issues in more detail below, and later, we will also examine the implications that control system response times, launch and recovery needs, and navigation requirements have on the software design of an autonomous system.

Finally, we note that Table 2.1 does not include all the capabilities needed for operation in high-threat and contested environments. It also does not include all the sensor and data-processing capabilities needed for specific missions (e.g., target detection or weapons control). We discuss some of these potential new UxS capabilities in the next section.

UxS in Future Operations

For more than a decade, U.S. forces have been engaged in combat against adversaries with limited technological capabilities. U.S. forces have focused on counterinsurgency and counterterrorism operations and nation-building efforts to help stabilize such countries as Iraq and Afghanistan and other areas in the Middle East and Africa. To support these efforts, unmanned systems have played a critical role on the ground and in the air, and as it pertains to airborne operations, U.S. forces have enjoyed air superiority in almost all areas of operation (with one possible exception, described below).

Recent military history will tell a different story about operations on the ground. In ground operations, U.S. forces have confronted adversaries that have employed difficult-to-counter asymmetric warfare tactics, such as using hidden IEDs to attack ground forces when they traverse unfamiliar, complex, rural, or urban terrain. One challenge of recent operations has to do with the large area that a limited number of U.S. ground forces must cover and secure in recent counterinsurgency operations. Because of this, in many areas of operation in Iraq and Afghanistan, U.S. ground forces have not enjoyed the equivalent of air superiority on the ground.

Although unmanned systems have been very useful in conducting wide area surveillance and targeting operations from the air, they have not been used to the same extent to do similar area surveillance tasks directly on the ground. One challenge with using UGVs is their relatively low speed. Another is that they have to move over complex terrain. Remote vehicle operators can guide UGVs moving over complex training terrain; however, there is an inherent tradeoff in the communications bandwidth available and whether it is used for surveillance and targeting or for terrain maneuver (i.e., driving) tasks. Also, human targets of interest in counterinsurgency operations are often very challenging to identify and differentiate from civilians. To our knowledge, the automatic identification of human individuals by a robot has not yet been demonstrated and would likely require significantly more autonomy than even advanced robots have today.[2] In contrast, ground robots have been used predominantly in explosive ordnance disposal (EOD) missions, which can be conducted effectively at shorter ranges where effective teleoperation is possible.

U.S. efforts to gain and maintain air superiority in future conflicts could be challenged by a competent adversary. Potential adversaries are developing capabilities to deny the United States air superiority should conflict arise or to make the cost of achieving air superiority very high. Consequently, the United States may face more demanding operational environments and adversaries with weapons able to threaten UAS operating at medium and high altitudes.

Of course, the United States has had significant success in developing manned aircraft with stealth capabilities to survive in such environments. However, stealth aircraft are very expensive and, as a result, there is an interest in exploring alternatives to degrading enemy air defenses without relying entirely on manned stealth aircraft.

This naturally raises the question as to whether unmanned systems can contribute to the task of achieving air superiority in A2/AD environments. Unmanned systems already exist that support the suppression of enemy air defense and the destruction of enemy air defense missions.[3] Some of these current systems are programmable.[4] Suppose they possessed greater autonomy to adjust their flight, surveillance, and targeting

---

[2]   The science of biometrics has advanced significantly in the past decade. Humans can now be identified by fingerprint-matching, iris scans, and facial recognition (with some error rate). The accuracy of biometric matching systems has improved significantly if high-quality samples are collected. However, biometric samples are currently collected using a partially manual process with human supervision and control.

[3]   Raytheon Company, "Miniature Air Launched Decoy (MALD)," undated.

[4]   Raytheon Company, undated.

plans. Inserting greater autonomy into such unmanned systems is one approach for making them more effective in A2/AD environments. Another approach is to make them remotely piloted or controlled. But this presents other challenges because adversaries may be able to jam or interfere with their wireless radio or satellite communication links. Generally speaking, high-capacity wireless communications links are relatively easy to jam because of the limited link margin such links typically have.[5]

The 2011 Iranian capture of the U.S. RQ-170 Sentinel, a stealthy unmanned surveillance aircraft, is one example of the possible challenges the United States may face in future conflicts. Open source reports can still only speculate on what exactly transpired during this event. However, reports indicate that the Sentinel was used to conduct surveillance of Iran's military and nuclear weapons development programs and, at some point during this mission, the RQ-170 was captured by Iranian forces. Some initial reports speculated that the RQ-170 was shot down, but footage of the system after the capture indicated that this may have not been the case, given that the airframe was intact and there was no visible crash damage, except possibly of the landing gear or bottom portions of the fuselage (which were hidden in photos released by the Iranian government). This has led many to question how the RQ-170 was actually captured and may indicate that the Iranians somehow took control of the vehicle and intentionally landed or crash landed it at an Iranian airbase. Did the RQ-170 malfunction as a result of a lost command link? Were the Iranians able to jam the control signal, spoof the aircraft's GPS receiver, or take over the control link? Doubt still remains as to whether Iranians have the wherewithal to be successful in such a jamming attack.[6] Nevertheless, the possibility of an adversary using electronic warfare to override U.S. controls and take control of the system is a concern for such high-value UAS as the RQ-170. This unfortunate episode raises the question of whether this UAS capture could have been prevented if the vehicle had been equipped with some limited autonomy that could have recognized the transmission of commands to land or fly to an Iranian airbase or that could have compared its planned flight path to new spoofed GPS coordinates it had just received.

Maritime navigation and surveillance missions share some of the characteristics of surveillance missions conducted from the air, but they also possess their own unique characteristics. Naval vessels move much slower than aircraft, so if they are discovered they are easier to track. However, the surveillance problem is more challenging. The surveillance horizon in maritime operations is much shorter than that in air surveillance. This is simply because low-altitude sensors have much shorter ranges regardless

---

[5]   This is not generally true of optical laser communication links. However, laser communications has its own drawbacks. One of these is environmental, as laser communication links can be disrupted or attenuated significantly in the presence of clouds or water vapor. They also present other cost and technical challenges, which are being addressed by the research and development communities. For these reasons, we do not assume that laser communications can be used to address U.S. communication needs in the near term.

[6]   David Axe, "Did Iran Capture a U.S. Stealth Drone Intact?" *Wired* (online), December 4, 2011.

of the power levels they may have (targets are occluded by the horizon). In addition, the maritime surveillance mission is at least as daunting as the air surveillance mission in that the size of potential theaters of operation are very large.

Also, just as the task of achieving air superiority will become more challenging in the future, so will the task of achieving maritime superiority. Adversaries are working on new antiship weapons and surveillance capabilities to find and target U.S. Navy ships. Therefore, the Navy shares a similar problem as the Air Force in some future conflict scenarios—having to operate effectively in A2/AD environments.

If USVs were able to operate effectively in this environment or contribute to the mission of degrading adversary capabilities, they could potentially be very valuable in future operations. Just as with unmanned aircraft, USVs face a similar conundrum in that to operate in an A2/AD environment, they have to be given either greater communications capabilities or greater autonomy.

Related RAND research for the U.S. Navy found that

> [A]dvances in autonomy and assured communications are path-critical for USVs to conduct complex missions and/or to operate in complex environments. Autonomy, assured communications, and mission or environmental complexity form a trade space. As environments or missions grow more complex, increasingly advanced autonomy and/or assured communications are required.[7]

And in the case of USVs, the option of providing greater communications capabilities appears to be even less attractive than in the unmanned aircraft case.

UxS Interoperability Challenges

Short-Term Interoperability Challenges

A number of UAS interoperability shortcomings have been observed in recent operations that hamper U.S. forces. These issues can make it difficult to share control of UAS with authorized but unanticipated tactical users. In some situations, other interoperability problems can prevent ISR data collected by one particular type of UAS to be shared with other tactical users that need the ISR data but that are equipped to use another kind of UAS ground control station (GCS). An immediate goal of the DoD is to resolve the UAS interoperability challenges identified in recent operations.

Some of these interoperability challenges result because UAS produced by one particular manufacturer can be controlled only by a GCS developed by that same manufacturer. Other interoperability issues are caused by incompatible communications equipment used on different UAS. Such interoperability issues may prevent the transfer of command and control functions from one military unit to another. It can also prevent or restrict the dissemination of UAS ISR data to military units that could

---

[7]   Savitz et al., 2013.

benefit from such data. Communications incompatibilities can be due to the use of different noninteroperable waveforms, the inability of certain UAS payloads to operate in specific frequency bands, the inability of UAS payloads to cooperate in or share the same frequency bands without interfering with each other, among other reasons.

A major task in resolving UAS interoperability issues is to precisely define and understand the technical basis or source of interoperability problems between systems and ground stations used by different military units. It can be difficult to troubleshoot and pinpoint the source of interoperability problems, especially before they occur on the battlefield. To do this, precise and accurate technical information is required for each system under consideration. The UAS I-IPT Task Force has found that current DoDAF architecture products produced by UAS program offices have not contained the information needed for interoperability analysis, in particular, the information needed for joint interoperability analysis. In other words, because of technical complexity, the required architecture products that acquisition programs produce do not contain the information needed to predict where and when interoperability problems will occur. There are several reasons for this acquisition process shortcoming, which we will investigate further below. These observations suggest that improved architecture products from acquisition programs are needed to better predict interoperability problems before they arise in the field.

Long-Term Interoperability Challenges

The interoperability shortcomings of current systems prevent full exploitation of the valuable information UAS are collecting today. Future concepts for the use of unmanned systems in joint operations include the possibility that air, surface, and ground unmanned systems would operate and cooperate together to contribute to or accomplish a growing range of military missions. To enable this degree of cooperation and synchronization, an even greater degree of information-sharing will be needed between unmanned platforms and the humans who control these systems. In addition, future unmanned systems will likely be developed with more autonomous capabilities. These unmanned platforms will have to be able to communicate directly with one another without human intervention to accomplish missions cooperatively. In other words, the interoperability needs of future unmanned systems are likely to be even greater than they are today.

Another interoperability goal of the DoD is to enable direct interoperability between an unmanned system and a manned system, even when these systems operate in different domains (air, surface, subsurface, and ground).[8] In addition, the DoD also aims to enable unmanned aircraft operating in the same area to share and process situational awareness and targeting information.

---

[8]   Department of Defense, 2011.

CHAPTER THREE

UAS Interoperability Initiatives Led by the Office of the Secretary of Defense

The OSD has undertaken a number of initiatives to improve UAS interoperability. This work is being spearheaded by the UAS I-IPT. In this chapter we survey the goals, activities, and products of the I-IPT.

UAS I-IPT Goals

The goals of the UAS I-IPT task force are to

1. develop a joint UAS interoperability plan that defines core capabilities and interfaces/functions to improve interoperability across DoD unmanned aircraft systems
2. recommend standardized interoperability implementations for UAS programs
3. develop and standardize the overall UAS architecture
4. establish an enduring process for interoperability through the development and sustainment of standards, interfaces, and protocols.[1]

The IPT is charged with developing an enduring interoperability improvement process in collaboration with OSD, the Joint Staff, the military services, combatant commands (COCOMs), and applicable interagency organizations.[2]

The DoD has had long-standing goals and plans to improve information-sharing by using network-centric concepts of operation and by making systems interoperable with the DoD global information grid (GIG). Another goal of the UAS I-IPT Task Force is to develop information standards to make UAS more net-centric (i.e., to make UAS information discoverable and shareable in the GIG).

To accomplish these goals, the I-IPT has established working groups that include representatives from each of the military services.

---

[1] Department of Defense, 2011.

[2] See Department of Defense, Interoperability–Integrated Product Team, "Mission Statement," 2010.

UAS Interoperability Profiles Working Group

The UAS interoperability profiles (USIPs) working group develops USIPs, which are sets of self-consistent interoperability standards that can be used to establish interoperability between aircraft and ground control stations and communication networks. However, it should be noted that although interoperability profiles are necessary, they are not sufficient for ensuring interoperability between different UAS system components. In addition, it is not clear that currently deployed UAS have to comply with recently established USIPs. Therefore, as with many DoD information system standards, it will likely be some time before they have a positive influence and effect on the interoperability of deployed DoD systems. Nevertheless, they represent a starting point from which new unmanned system acquisition programs can and should start.

Mission Integration Working Group

The mission integration working group is focusing on information-sharing across service, combatant command, and processing exploitation and dissemination (PED) system boundaries. It has been active in demonstrating the ability to exchange information between systems and organizations that use different UAS or different UAS ground control or related information-processing systems. Many of these initiatives have very specific information-sharing objectives. They represent the "picking of low hanging fruit" for improving UAS interoperability with the limited resources currently available from existing programs of record (PORs) and related joint capability technology demonstrations. Nevertheless, this important working group activity can help bridge specific UAS interoperability gaps between the operational and intelligence communities.

Horizontal Integration Working Group

The horizontal integration working group developed an "as-is" joint common UAS architecture (JCUA) in 2012 and is currently working on a "to-be" joint UAS architecture. The JCUA includes operational views, system views, and a set of common terms for defining UAS architectures. One key task this working group accomplished in developing this as-is architecture was to review the architectural products of individual UAS programs of record. The as-is architecture reflects the current status of interoperability, or lack thereof, between unmanned systems being used by the different services and combatant commands.

UAS Control Segment Working Group

The UAS control segment (UCS) working group has developed a UCS architecture. This is a service-oriented architecture (SOA) for a common UAS GCS, which is designed to enable a UCS-compliant GCS to interoperate with and control a wide range of DoD UAS. This open architecture approach allows industry to compete openly and without restrictions imposed by proprietary systems.[3] To demonstrate the maturity of this architecture approach, the UCS working group funded product development efforts to demonstrate the use of the UCS architecture and illustrate its potential for joint interoperability and integration of UAV and GCSs. The UCS working group is now refining this architecture by more precisely specifying its software standards. Another objective of the UCS architecture is to enable software reuse and to maximize the use of commercial off-the-shelf technology. Shown in Figure 3.1 is the UCS technical reference model for the UCS architecture.

It is important to point out that the UCS technical reference model (TRM) supports a modular architectural approach that the authors believe to be useful in supporting innovative software developments for UAVs and GCS. This point is supported by the fact that the UCS has been used to develop a UAS "App Store," which is now being used to develop UAS-related software applications and to share such applications among UAS developers and software developers.

Figure 3.1
UCS Technical Reference Model

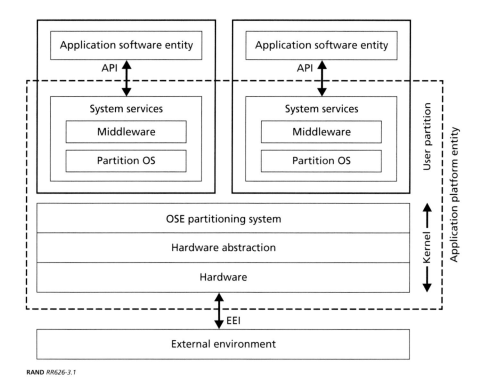

RAND RR626-3.1

---

3   The U.S. Navy, in particular, expressed a desire to migrate to this common architecture approach. For more background on UCS working group efforts, see Edward Lundquist, "Open Business Model," *Seapower Magazine*, November 2013.

Review of UAS I-IPT Working Group Products

USIPs

The standards development activities of the USIP working group are a necessary process of the DoD acquisition community, but, as noted, this step is not sufficient for establishing UAS interoperability. USIPs have to be adopted by DoD UAS acquisition PORs and incorporated into UAVs and GCS. USIP adoption plans by PORs are currently not tracked by the I-IPT. We believe that USIP tracking should be done by the I-IPT. It could provide valuable information regarding the utility and potential shortcomings of USIPs.

JCUA

The horizontal integration working group discovered an important limitation of the architecture products produced by UAS acquisition programs. They found that the existing DoDAF architecture projects produced by UAS acquisition programs are not suitable for interoperability analysis. To address this shortcoming, this working group has developed a common architecture that describes all DoD UAS with consistent terminology and consistent graphical formats—the JCUA.

One challenge found in developing the as-is JCUA is that UAS components and functions are labeled in different ways in the architectures of different UAS PORs. This means that there are significant inconsistencies in the semantics or terminology used by the different PORs. Because of this inconsistency, it is difficult to tell where the architecture views of different programs differ and where they align. Although a "common" taxonomy has been created and is used in the as-is JCUA for key UAS systems, components, and functions, it has not been adopted by UAS PORs, and so it does not map easily to the taxonomy used in current UAS architectures.

The to-be JCUA that the horizontal working group is developing will represent a target architecture that UAS programs of record will be asked to converge to. It will also incorporate the USIPs developed by the USIP working group. Just as with USIPs, at present, JCUA adoption is voluntary by UAS PORs.

UCS Architecture

The UCS architecture uses USIPs developed by the USIP working group. The UCS architecture appears to be the most mature product yet developed by the UAS I-IPT. It has been used to develop a prototype GCS that can control current UAS from multiple PORs.

Aligning Future UAS Architectures

The UAS I-IPT working groups offer two families of architecture products that UAS PORs are encouraged to adopt or converge to. One of these, the UCS, has demonstrated utility. In contrast, it is not clear whether the JCUA, as currently structured, provides the best approach for integrating the UAS programs of the military services and whether it can effectively address the UAS interoperability challenges described above.

One challenge with using architectures to improve interoperability is the expense involved in developing and modifying architectures, which would necessarily have to precede any actual changes to the systems themselves. One must also factor in the cost of implementing the changes in the actual systems. Each POR has already developed unique architecture and architecture views. Unless there is a major block upgrade to a UAS system, it is unlikely that these programs would invest the time and effort to develop new architectural views consistent with the JCUA. Furthermore, UAS PORs may have to incur additional costs to align their architecture products with two different target architectures (the JCUA and UCS architecture).

The DoD is aware of the high cost of developing DoD architecture products. It has revised architecture development guidance in an effort to reduce the cost of the development and of reviewing these architectural products, and to increase the utility of architecture products for interoperability analysis. A significant change in this guidance occurred in moving from DoDAF 1.5 to DoDAF 2.0. DoDAF 2.0 enables architectures to be federated if they follow the correct architecture format specified in the DoDAF 2.0 guidance and if the two contributing architectures use the same architecture metadata or semantics.

The JCUA may include too many design points that must be aligned and synchronized among the many UAS programs of record. Furthermore, it is unclear whether the JCUA is consistent with the UCS architecture. The I-IPT is working to develop a strategy to integrate these two architectures and to make sure that future versions are aligned. We will offer some recommendations for how this can be done.

CHAPTER FOUR

Unmanned System Autonomy: Limitations and Opportunities

The discussion in the previous chapter indicated that developing unmanned systems with greater autonomy could provide significant operational benefits to all the military services. Greater autonomy could expand the types and complexity of missions that unmanned systems could perform. For example, greater autonomy in UAS could decrease the pressure on communications systems through onboard processing, and UGVs could provide support for more dangerous missions without putting U.S. forces at risk. Furthermore, with greater autonomy, unmanned systems operating in the air, ground, and maritime domains may be able to perform joint combat missions in contested or A2/AD environments.

However, achieving desired levels of autonomy may not be easy, depending on the type of unmanned system or the specific mission.[1] DoD acquisition programs to date have so far had limited success in fielding UxS with autonomous capabilities. This systems engineering challenge has prompted OSD leaders to task the DSB with studying this issue and recommending ways to address this shortcoming.

The DSB found that misperceptions about autonomy were limiting the development and adoption of autonomous capabilities in unmanned systems.[2] It noted that much past research centered on how to achieve fully autonomous unmanned systems and on defining levels of autonomy. The DSB found that past research on levels of autonomy had led to unproductive results and that, instead, a more focused capability-based approach to research and development was needed. In addition, it also found that research was not well coordinated among the different parts of the research establishment. Finally, it also found that inserting autonomous capabilities into acquisition programs is a difficult process.

There are number of potential reasons besides those mentioned by the DSB as to why this has proven to be difficult. First, most UAVs have been developed by aircraft manufacturers who, although adept at developing aircraft and integrating traditional

---

[1] For example, for more background on the challenges of autonomy for UUVs, particularly for intelligence collection missions, see Robert W. Button, John Kamp, Thomas B. Curtin, and J. A. Dryden, *A Survey of Missions for Unmanned Undersea Vehicles,* Santa Monica, Calif.: RAND Corporation, MG-808-NAVY, 2009.

[2] Defense Science Board, 2012.

avionics devices into unmanned systems, are not necessarily leaders in developing new software-intensive systems or innovative software products. In this regard, it is interesting to note that the first driverless cars have been developed by Google and not by traditional automobile manufacturers. The second reason for the lack of progress in developing autonomous systems is that UxVs have tended to be developed by a single contractor using proprietary standards for key internal and external information and communications systems interfaces. The proprietary nature of these interfaces makes it harder to incorporate new software functionality or to upgrade the system using software from different vendors. In other words, the lack of an open or modular system architecture has hampered innovation in developing new software technologies for UxVs. It can be argued that for performance reasons, and to ensure the timely response of UxVs to telerobotic control, very tight integration is needed between UxV control surfaces, propulsion systems, and avionics. Running middleware software that may be needed to support communications in an open architecture system requires additional memory and microprocessor resources, which in systems with limitations in both areas may introduce time delays in UxV control loops. However, with the ever-increasing speed of microprocessors, embedded systems, and other microchips, it has become easier to introduce middleware into UxV architectures that abstracts communications interfaces and that, in turn, enables open or standardized module boundaries to be introduced.

UAS Autonomy

Researchers and industry firms have been experimenting with inserting limited autonomous capabilities into UAS for some time. A number of UAS autopilot products can perform autonomous landings at presurveyed airbases that are equipped with differential GPS guidance systems, even for small UAS.[3] In addition, in 2012, researchers demonstrated a UAS autonomous landing capability at a non-presurveyed site using only the UAS video cameras.[4] This system does require some information on the landing site and that it be preloaded into the aircraft before the mission.

Autonomous capabilities such as autopilot landing can reduce the number of personnel needed to operate a fleet of UAS. The U.S. Air Force Research Lab (AFRL) has been conducting research on UAS autonomy for many years. Autonomous cooperative UAS capabilities are assessed by AFRL to fill a critical Air Force need. Indeed, the Air Force UAS Flight Plan published in 2009 articulated a need to reduce the number of

---

[3]   Rockwell Collins, "Athena 311 Integrated Flight Control System," 2014; MicroPilot, "MicroPilot–Products–MP2128g," 2011.

[4]   Paul Williams and Michael Crump, "Intelligent Landing System for Landing UAVs at Unsurveyed Airfields," *Proceedings of the 28th International Congress of the Aeronautical Sciences,* 2012.

remote UAS pilots because the number of UAS orbits was growing rapidly. With the drawdown in current operations in Afghanistan, this pressing need has been somewhat reduced. However, it may emerge again in future conflicts; in particular, in a "high-volume fight," there may be little time and not enough remote pilots to micromanage a fleet of UAS, especially in a contested environment where command and control communication links may be disrupted. In this case, autonomous capabilities will be a critical enabler.

Specific AFRL UAS autonomy goals are given below:

- cooperative autonomous decisionmaking for teams of UAS in surveillance and tracking scenarios
- understanding the interplay of trajectory planning and mission planning layers
- increased consideration of operations in adversarial environments (lack of communication, active adversary).[5]

Increasing the endurance of UAS is a research area where greater UAS autonomy could potentially be beneficial. If UAS could be refueled in the air, their range and endurance could be increased significantly. A series of manned and unmanned programs have been examining this capability for over a decade.[6] The most recent and perhaps the most ambitious of these programs is the DARPA KQ–X program, which will demonstrate the ability of one Global Hawk UAS to refuel another at high altitude.[7] Such a capability requires very accurate autonomous navigation and flight control capabilities. Because of the time lags with satellite communication links, it is very difficult and risky for a UAS to be refueled in the air by remote control.

UAS have been used in place of manned systems to reduce the risk of casualties or the capture of aircraft pilots by an adversary. A semiautonomous, rotary-wing UAS has been developed by Lockheed Martin in collaboration with Kaman Aircraft.[8] This system is based on the Kaman Aircraft manned K-Max helicopter and can be remotely piloted or operated autonomously. This UAS has a 6,000 lb payload lift capability and is designed to provide logistic support to ground forces in high-threat environments. Two autonomous K-Max UAS have been operating in Afghanistan since 2010 to support U.S. Marine Corps operations, although one crashed in mid 2012.[9] It has been designed to land autonomously at austere bases and was landing when it crashed in Afghanistan. After this event, K-Max operations in Afghanistan were suspended. Since

[5]   Derek Kingston, "UAV Autonomy," September 19, 2012.

[6]   Rebecca Grant, "Refueling the RPAs," *Air Force Magazine,* March 2012.

[7]   Grant, 2012.

[8]   Lockheed-Martin, "K-MAX Lockheed Martin," undated.

[9]   "K-MAX Crashes on a Mission in Afghanistan," Defense Update, June 17, 2013.

that time, the Army has demonstrated K-Max autonomous landing capability using a high-resolution laser imaging detection and ranging (LIDAR) sensor.[10]

Another important area where autonomy would be beneficial is in the onboard processing (OBP) of ISR information collected by UAS. More-capable, higher-resolution, and wide-area surveillance sensors are being developed and deployed on UAS. One example is the Autonomous Real-Time Ground Ubiquitous Surveillance (ARGUS) imaging system (IS) sensor developed by DARPA. This sensor is actually an array of solid-state digital cameras that together can produce a high-resolution, wide-area image that is 1.8 gigapixels in size.[11] The system can also stream the video data and can stream over one million Terabytes of video per day. Needless to say, such a sensor would overwhelm the communications systems that are now used on even the most advanced UAS designed that collect ISR data. The only way the full take of the ARGUS sensor could be used is to store most of the data it collects onboard the vehicle or process it onboard using software algorithms to detect, identify, or track targets. This is an active area of research in the defense community. However, it is not clear whether a common approach has been developed that can then be used to transition autonomous target detection and tracking capabilities to the many ISR-capable UAS in the DoD fleet.

UMS Autonomy

In this section, we discuss both types of UMS: USVs and UUVs. First, we consider UUVs. An important consideration in the development of UUVs is autonomy. Achieving and maintaining communication with underwater vehicles and even with surface vehicles is technically challenging, especially at longer ranges. Water attenuates radio waves and other wireless signals that can easily be used at long range in air-to-ground or air-to-air communications. This means that high-bandwidth communications underwater is largely impractical using traditional communication technologies. Although there has been some experimentation with laser communications for underwater applications, laser communications systems are expensive and consume considerable amounts of power. Because of these communications limitations, UUVs that do not require continuous communications links are essential. For example, autonomous path planning is needed to avoid underwater obstacles and unanticipated terrain features.

USVs also will likely require autonomous capabilities if they are to perform demanding missions effectively. Although USVs can be remotely commanded using

---

[10] Graham Warwick, "Unmanned K-Max Gets Cleverer," *Aviation Week,* August 11, 2013.

[11] Paul Szoldra, "Drone Spying Capabilities Are About to Take Another Huge Leap," *Business Insider,* January 29, 2013.

satellite communications in some environments, such communication links may not be available in A2/AD environments. If such surface vehicles cannot be remotely controlled and if the environment contains natural hazards or faces potential threats from adversary surface vessels, the USV will have to autonomously maneuver to avoid these challenges. In addition, commercial ship traffic can be a hazard, even in peacetime under certain conditions (e.g., lost satellite link or unintentional interference). This implies that adaptive route planning and the ability of the vehicle to autonomously generate its own situational awareness information using multiple sensors and data fusion will be important capabilities for USVs as well. For both countermine and anti-submarine warfare, target detection and tracking are also important capabilities that may have to be accomplished autonomously by unmanned vehicles. Consequently, autonomous capabilities become much more important for USVs and UUVs if they are to have useful operational capabilities.

Another UAS research initiative that may increase the autonomous capabilities of UAS is the NASA Global Hawk autonomous refueling demonstration project.

The U.S. Navy's UUV systems have moderate levels of autonomy today. Examples of their autonomous capabilities are:

- GPS/Doppler-aided navigation
- autonomous path planning and execution based on onboard world map
- terrain-following, keep-out zone avoidance
- autonomous decisionmaking and cue generation for noncombat missions
- dynamic replanning based on sensor input (acoustic, radio frequency [RF], chemical, etc.), vehicle health, and mission objectives and priorities
- cross-deck advanced autonomy on multiple classes of vehicles—interface to various vehicle controllers and payload controllers.[12]

The Navy is working on developing higher levels of autonomy the Navy for UUVs. These include:

- long transit and autonomous planning and control to precise local insertion without GPS-aided navigation (i.e., bottom mapmatching/feature-based navigation)
- adaptive area surveys with automated target detection, classification, and recognition
- robust sense and avoidance of hard-to-image/classify obstacles
  - surface vessel detection and avoidance
  - fishing gear detection (based on cues)
  - threat avoidance (perception is the hard part)
  - RF spectrum threat counterdetection

---

[12] Captain Duane Ashton, "Unmanned Maritime Systems Autonomy," presentation delivered at the 10th International MIW Technology Symposium, PMS 406 Unmanned Maritime Systems, May 2012.

- autonomous sensor data fusion
- collaborative behaviors
- fault detection and response
- autonomous sensor reconfiguration to meet changing mission needs.[13]

The Navy envisions that even higher levels of autonomy will be needed in future operations. These more sophisticated forms of autonomy will probably be achieved only in the long-term:

- fishing net detection, avoidance, and extraction
- counterdetection awareness and response
- dynamic threat perception and adversary intent
- autonomous decisionmaking to support use of weapons
- advanced collaborative behaviors
- survivability for long-duration, complex missions.[14]

UGS Autonomy

DARPA Initiatives

There are many reasons to pursue the development of autonomous ground vehicles for military applications. Ground combat is fraught with risk. Ground combat vehicles are vulnerable not only to adversary armored forces but potentially also to infantry forces equipped with advanced antiarmor weapons. DARPA made a significant leap forward in the development of autonomous ground vehicles when it held the first DARPA grand challenge in 2004.[15] This was essentially a race in a rural environment where competing teams attempted to navigate a racecourse filled with a variety of obstacles reminiscent of realistic conditions and reach the finish line before the other competitors. In the first grand challenge, there was no winner, as no team was able to completely navigate the course successfully.[16] A second DARPA grand challenge was held the following year in 2005. This race also took place in a rural environment in which the race competitors were isolated from other vehicle traffic. In the 2005 race, an autonomous ground vehicle development team from Stanford University was declared the winner. The next grand challenge race took place in 2007. This racecourse was substantially different from the first and was designed to be reminiscent of urban terrain. A team from Carnegie Mellon University won this race, with an average speed of 14

---

[13] Ashton, 2012.

[14] Ashton, 2012.

[15] DARPA, "Grand Challenge Overview," March 2004.

[16] DARPA, 2004.

km/h. Just as in the previous grand challenges, there were no civilian ground vehicle traffic on the race course. The only time unmanned ground vehicles came into contact was when competitors attempted to pass one another.[17]

Army Initiatives

The Army has developed a roadmap for unmanned ground systems.[18] The roadmap briefly discusses the autonomy needs of future UGS and some of the programs that have been sponsored by the Army to demonstrate such autonomous capabilities. UGS autonomy needs fall into two general categories: object recognition and intelligent navigation.

The Army recognizes three autonomy needs in object recognition:

- recognize combatants/noncombatants
- recognize other living entities
- recognize vehicles, roads, paths, and markers.[19]

Three autonomy needs are also identified in the intelligent navigation category:

- avoid static and dynamic obstacles
- predict motion of dynamic objects
- obey traffic regulations as appropriate.[20]

Google and Automaker Initiatives

Autonomous ground vehicles hold the promise of not only reducing the number of soldiers in harm's way but also offering the potential of making automobile travel more productive for civilian car owners and truck fleet operators. Some observers have claimed that autonomous vehicles may reduce traffic congestion on highways by making traffic flow more efficient. Although this claim has yet to be substantiated, research and development on autonomous vehicles has increased substantially in just the last few years. Google is arguably the leader in the development of autonomous vehicles and has a fleet of such vehicles that use technologies first employed in the DARPA grand challenge series races. The DARPA autonomous vehicle fleet has logged over 200,000 miles with only a few accidents.[21] Currently, every major automaker is

---

[17] "DARPA Grand Challenge," *Wikipedia,*" undated.

[18] Robotic Systems Joint Project Office, "Unmanned Ground Systems Roadmap," July 2011a.

[19] Robotic Systems Joint Project Office, 2011a.

[20] Robotic Systems Joint Project Office, 2011a.

[21] Erico Guizzo, "How Google's Self-Driving Car Works," *IEEE Spectrum,* October 2011.

developing autonomous vehicle technologies.[22] So far, it appears that these developments by independent automakers rely on their own proprietary software architectures. Google has also kept its autonomous vehicle software architecture proprietary, in contrast to its approach with mobile phone software, where it made Android available as open source software to all mobile phone developers.

The needs and objectives of commercial autonomous vehicles mirror but are not identical to those expressed by the Army in its autonomous ground vehicle roadmap. Civilian autonomous vehicles must be able to recognize the presence of pedestrians and other vehicles to avoid collisions. They must also be able to recognize the boundaries of roadways and highways and to obey traffic laws, road signs, and traffic lights. Google autonomous cars address these navigation and obstacle-avoidance challenges by making use of GPS and roadway information furnished by Google maps. They also employ LIDAR sensors, which generate a real-time map of the local environment surrounding the car. Each Google car is loaded with much more than Google maps, however, before it embarks on its trip. It is loaded with a high-resolution, three-dimensional map of the planned route. The on-board autonomous system then forms a difference map of the anticipated or a priori three-dimensional picture with the three-dimensional picture constructed using its own LIDAR sensor at many points along the driving route. Using this approach, the system can detect the presence of pedestrians or other vehicles, and it uses this information to dynamically adjust its planned route.[23]

It should be noted that the Google approach to autonomous vehicles is very data-intensive. This data-intensive approach can potentially work for civilian vehicles that operate on roads and highways mapped by Google. However, it would not necessarily work for combat vehicles that have to operate in rural terrain for which high-resolution maps may not be available. This raises the question of whether there may be more appropriate approaches to combat UGV autonomy. On the other hand, it appears that civilian autonomous vehicles are relatively close to deployment by commercial automakers. The chief executive officer of Nissan, for example, has stated that his car company would have an autonomous vehicle ready for the market by 2020.[24] The amount of research and development funds going into civilian autonomous vehicle development will likely greatly exceed that available for UGV R&D in the DoD budget over the next decade. This raises the question of whether the DoD should work with one or more leading civilian automakers and adapt their emerging software architectures for autonomous vehicles.

---

[22] James M. Anderson, Nidhi Kalra, Karlyn D. Stanley, Paul Sorensen, Constantine Samaras, and Oluwatobi A. Oluwatola, *Autonomous Vehicle Technology: A Guide for Policymakers,* Santa Monica, Calif.: RAND Corporation, RR-443-1-RC, 2014.

[23] Guizzo, 2011.

[24] Damon Lavrinc, "Nissan Promises to Deliver Autonomous Car by 2020," Autopia, *Wired.com,* August 27, 2013.

CHAPTER FIVE

UxS Architecture Developments

In the past, the DoD encouraged the use of a MOSA approach in the acquisition of DoD systems. Yet, although there are multiple reasons to use such an approach, there are also a number of reasons why contractors are wary of adopting it. In any acquisition program, there will be tensions between the government and contractors about how best to build and design a system to meet the operational needs of the warfighter in a cost-effective manner. In this chapter, we review the principles of open system architecture and discuss some of these tensions and how they apply to the development of unmanned systems.

As described in the DoD open system architecture (OSA) guidebook, there are five fundamental principles in this approach:

1. modular designs based on standards, with loose coupling and high cohesion, that allow for independent acquisition of system components
2. enterprise investment strategies, based on collaboration and trust, that maximize reuse of proven hardware system designs and ensure that we spend the least to get the best
3. transformation of the life-cycle sustainment strategies for software-intensive systems through proven technology insertion and software product upgrade techniques
4. dramatically lower development risk through transparency of system designs, continuous design disclosure, and government, academia, and industry peer reviews
5. strategic use of data rights to ensure a level competitive playing field and access to alternative solutions and sources, across the life cycle.[1]

The guidebook recommends that OSA technical requirements be based to the maximum extent possible on open standards and, when no such standards exist, for the acquisition program or the government to create them to support the adoption of

---

[1]   Department of Defense, "DoD Open Systems Architecture (OSA) Contract Guidebook for Program Managers," June 2013.

the OSA methodology. We believe that it is unlikely (for the reasons given below) that if open standards do not exist for a particular part of an unmanned system, industry will create such standards independently. Therefore, it may be necessary for the government to take the lead in developing such open standards.

Until recently, the government has not taken a leading role developing such standards for UAS. To date, government-led efforts have focused predominantly on the ground control segment of UAS. In contrast, UAV prime contractors have typically had complete control over the standards used internally on the aircraft platform.

An open system architecture would be beneficial in the development of unmanned systems because it would potentially enable unmanned systems to use payloads developed by different contractors. These payloads could then be swapped out for particular types of missions or for use in specific operating environments (e.g., one particular payload may operate in a spectrum band that would not cause frequency interference in some environments, whereas another payload, such as a communications data link or a radar, could operate in a frequency band that is vacant in a particular region of the world). In addition, as discussed below, such an approach could also enable advanced capabilities to be integrated into unmanned systems that already exist. It may enable cost-effective software upgrades to existing systems that could increase the autonomous capabilities of the unmanned platform.

One could argue, and undoubtedly some defense contractors will argue, that such an open approach is not necessary and that prime vehicle contractors for current unmanned platforms could independently develop the new autonomous capabilities needed to support operations in more complex environments or missions. However, recent developments in the commercial world did not support this view. The first driverless cars with autonomous navigation and route-control capabilities were developed by a leading software company (Google) and not by a leading automobile manufacturer. Although many automobile manufacturers are now working on driverless cars, it appears that a company with unique expertise in software development has taken the lead in developing such autonomous capabilities. Therefore, one can surmise that a similar type of market advantage may apply to the development of autonomous capabilities for unmanned military systems.

Nevertheless, it may be unrealistic to advocate for and believe that a complete open system architecture approach should be taken for the development of next-generation unmanned systems. We believe that such an approach is unrealistic for one important reason—the intellectual property held by DoD contractors. Intellectual property is an important motivating factor in the behavior of private-sector and DoD firms. Although use of open source software is significant in industry, in most systems, a major part of the software code base is still considered proprietary. In addition, many unmanned systems used by the DoD were originally developed by small, innovative companies. These unmanned systems work well because of the tight integration between hardware and software that enables the effective remote control or piloting of such unmanned systems. This tight integration is accomplished by the use of proprietary interfaces and software code bases, which are the intellectual property of these companies and contractors. They will be reluctant to provide these to the government, as they may lose a competitive advantage in the marketplace if they do so. In other words, the release of this technical information would enable contractors to copy some of their innovations and produce competing products for the DoD. These factors lead to some of the tensions between industry and the DoD acquisition community alluded to above.

These factors also likely contribute to reasons why the adoption of an open system architecture approach has been particularly slow in the unmanned system industry and in DoD acquisition programs. Therefore, in this report, we do not advocate a complete open system architecture approach for unmanned systems. It would be unrealistic and counter to the interests and self-preservation instincts of private-sector firms. Nevertheless, important mission needs should be met by the next generation of unmanned systems, and we believe that this can be done by adopting the partially open systems architecture (POSA) approach for unmanned systems.

By a partially open systems architecture approach, we mean that only some of the key internal interfaces are defined using open standards and are published and made available to authorized contractors. In addition, in such an approach, unmanned system programs should be defined and structured to contain at least three modules that would provide essential building blocks for the POSA. One of these modules would include the external communications payload and its interfaces to the rest of the vehicle. An example of the type of open standards that can be used to define POSA communications interfaces are video standards adopted by the UAS-I IPT. In many cases, such standards are based on commercial standards, such as those established by the motion imagery standard board, for example.[2] The second module would include an autonomous software-based control system and command interfaces to both remote ground control stations and the key flight and engine control systems of the aircraft, such as the navigation systems, autopilot, and control surfaces. In addition, another possible module in the POSA would include the primary mission payloads (sensor and weapon payloads). The sensor module component of the POSA would define application layer interfaces for sensor data and also possibly for environmental data that would be used in software algorithms that perform OBP to update vehicle flight paths or routes or to identify and track targets in a sensor field of regard. Examples of application data standards for the former application include KML standards for geospatial data. Similar open systems and federated architectures are now widely used by such technology firms as Google, Microsoft, and ESRI for PED functions in intelligence systems, which could have significant roles to play in supporting the functions onboard

---

[2]   Amado Cordova, Lindsay Millard, Lance Menthe, Robert A. Guffey, and Carl A. Rhodes, *Motion Imagery Processing and Exploitation (MIPE),* Santa Monica, Calif.: RAND Corporation, RR-154-AF, 2013.

autonomous UxS. These systems are supplanting many of the older, highly proprietary, expensive, turnkey systems originally produced by defense contractors Northrop Grumman, Lockheed-Martin, and others that provide PED functionality. Google, Microsoft, and ESRI also all produce development kits to ensure interoperability, and they test new systems before they allow them to interface with their databases.[3] POSA standards for sensor or navigation modules could also use some of the same standards for communications interfaces as those used to define the POSA communications module.

What should be the scope of an unmanned systems architecture based on a POSA framework? For example, is it possible for a single architecture to apply and be used by system developers of UGS, UAS, and UMS? Above, we described the significant differences between unmanned systems that operate in different environments. System response times, platform shape and control surfaces, and other aspects of the systems are significantly different, as described in Table 2.1. For these reasons, we do not believe that a single POSA should be developed that applies to all types of unmanned systems. The architecture would likely be unwieldy, be too complex, and have too many caveats associated with it to provide a useful engineering design tool or be useful in interoperability and subsystem integration analysis.

We also note that there can be significant differences in the design and functionality of small and large UAS. Large UAS may use jet engines for propulsion and require the ability to operate at long ranges with satellite communication links. In contrast, very small or micro UAS would operate at much shorter distances from their home station or controllers, and would have much smaller payloads. They would not have the payload capacity or need for satellite communication links. Therefore, small and large UAS internal communications interfaces would likely be substantially different. We envision a family of POSAs for the wide variety of unmanned systems used by U.S. military forces.

Some critical open system architecture concepts are needed to support a POSA for unmanned systems. We believe that there are two key aspects of unmanned system design that need to be included in the architecture specification. The first is a design principle to separate system hardware from software using a middleware layer to the software architecture. This can be done in many ways and so it is important to consult with industry on what is considered to be the most technically efficient way to introduce middleware into a real-time control system that would be needed for an unmanned system. Clearly, for such an application, enterprise middleware products would not be appropriate. At the same time, a custom-designed Java-based middleware messaging system would not have sufficient openness to meet all the needs of the open system architecture approach for which we advocate. There is one leading

---

[3]   Google Developers, "KML Tutorial–Keyhole Markup Language," November 14, 2013. See also "Esri Geoportal Server | Open Source Metadata Management," undated.

open source middleware set of standards that can be used and that is already in use in unmanned systems and can form a key component of POSA. The first is the Data Distribution Service (DDS) set of standards that is managed by the open management group (OMG). DDS has a number of desirable technical characteristics for use in real-time systems and real-time control problems. It has demonstrated very low latency or time delay and message delivery between DDS nodes. It can also be implemented without the use of intermediate-level nodes or servers, which reduces system requirements and complexity. DDS has already been adopted and incorporated into the UAS I–IPT common grounds control system standard. The second key set of standards that need to be defined concern message command metadata. Contractors of current UAS need to disclose the full semantics of the command sets used by their autopilots and flight control systems. The semantics need to be incorporated in the POSA to enable them to be shared with contractors developing autonomous capabilities. At a minimum, technical standards and related specifications, requirements, source code, metadata, interface control documents, and any other implementation and design artifacts that are necessary for a qualified contractor to successfully perform development or maintenance work for the government should be made available throughout the life cycle. As described above, OSD (AT&L) has spearheaded the development of a joint common architecture for UAS (JCUA). However, the JCUA does not explicitly incorporate modular open system architecture principles. In fact, another architecture development effort spearheaded by OSD (AT&L)—the common ground control system architecture for UAS—does contain a modular structure and is informed by MOSA principles.

Below, we review industry and government initiatives for the development of modular open architectures for unmanned systems. First, we consider such developments for UGVs, where there has been significant progress in the last few years. Then, we address similar efforts for USVs and UUVs. For the latter unmanned systems, we find that although efforts are not as far along toward developing a MOSA approach for unmanned maritime vehicles (UMVs), significant progress has been made recently. Finally, we consider developments with UAS. We review some of the past efforts and more recent efforts toward developing a modular open system architecture for unmanned aircraft systems.

UGV Architecture–Related Developments

Unmanned ground systems developed in the 1990s possessed many of the interoperability problems that have already been discussed for UAS; in particular, the inability of operator control systems made by different vendors from being able to control UGVs from different vendors. This challenge, among others, has led to a desire to develop a platform-independent architecture that can enable components from different vendors to be used together in the same robotic ground system. This would enable configurable payloads, some measure of hardware independence, and the ability to insert new technologies and software upgrades into the system over time.

Joint Autonomous Unmanned System Architecture

In the mid-1990s, the DoD spearheaded the development of the joint autonomous unmanned system (JAUS) architecture. JAUS is an open and scalable service-based architecture designed to support unmanned vehicle developments that are platform-independent and hardware-independent and that can be used across a number of missions. It also defines a service's communication vocabulary—the JAUS service interface definition language. This provides messaging semantics and, specifically, message headers that can be used by UGV system components to translate into component and subsystem messages.

Industry plays an important role in the development of the JAUS architecture. The architecture is now maintained by the AS4 Unmanned Systems Technical Committee of the Society of Automotive Engineers (SAE). Over time, the architecture and its associated standards have been more precisely defined to reduce ambiguities. Key components of the architecture include the JAUS transport standard, which defines communications packet standards and address headers for Transmission Control Protocol, User Datagram Protocol and serial communications links. Another key element is the JAUS core service set, which establishes a common set of services for distributed systems. A third key element is the JAUS mobility service, which migrates mobility related components from the JAUS reference architecture (an earlier version of JAUS) to the new SAE JAUS architecture standard.

The top level structure of the JAUS architecture is illustrated in Figure 5.1.

In the JAUS technical reference model, the top level of the architecture is the system, which encompasses a collection of subsystems. The system can include one or more UGVs as subsystems and one or more Operator Control Units (OCU) as subsystems. A node is any physical computing (PC) endpoint in an UGV or OCU that has a physical address where messages can be sent from or received (e.g., an IP or serial link address). For example, a node might be a computer or microcontroller within a subsystem. Nodes can host one or more components, which can be applications or threads running on the node. The lowest level of the TRM is the component. Examples are shown in the figure. Components can use one or more services in the JAUS SOA. Subsystems, nodes, and components can communicate using JAUS messages.

The U.S. Army has continued to use JAUS to improve UGV interoperability. The DoD robotic systems (RS) Joint Program Office (JPO) has developed interoperability profiles (IOPs) for military UGVs that are based on the JAUS.[4] Elements of the JAUS UGV IOP define new nodes in the top-level JAUS technical reference model. Figure 5.2 provides an example of some of the nodes that can be defined. In the example, the

---

[4]   Robotic Systems Joint Project Office, 2011a.

Figure 5.1
JAUS Technical Reference Model

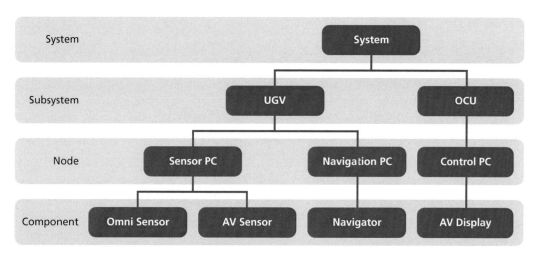

SOURCE: Daniel Barber, "Joint Architecture for Unmanned Systems," University of Central Florida, Institute for Simulation and Training, January 29, 2011.
RAND *RR626-5.1*

Figure 5.2
Example JAUS UGV IOP Reference Model

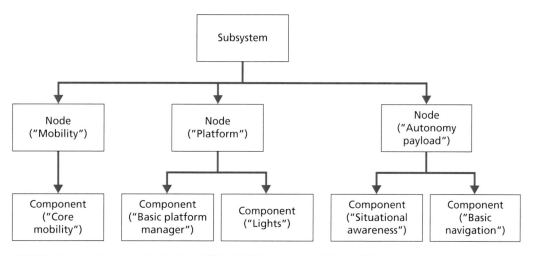

SOURCE: Robotic Systems Joint Project Office, "UGV Interoperability Profile (IOP)—Overarching Profile JAUS Profiling Rules," Version 0, SFAE-GCS-UGV MS 266, December 21, 2011b.
RAND *RR626-5.2*

"lights" component refers to head- or taillights of the UGV (which can be controlled either remotely or by means of an autonomous decisionmaking engine on the vehicle). Of particular interest to this study is that an autonomy node can be defined in the

UGV IOP. The UGV IOP also provides basic interoperability standards for leader-follower UGV operations and to dynamically redirect video streaming packets to a new endpoint in the JAUS network.

The robotic systems JPO UGV IOP is envisioned to apply to a range of vehicles. The applicable Army UGV class of vehicles (CoVs) are listed below:

- Warfighter transportable CoV is the UGV class small enough for warfighters to carry for extended periods. Within this class are the single warfighter and crew served robotic systems.
- Vehicle transportable CoV is larger than soldier transportable CoV and must be transported by another system, such as a truck or a trailer or towed to its mission location.
- Self transportable CoV is the UGV class large enough to transport itself and required payloads for extended periods.

The applique system is an add-on robotics conversion applique kit that will enable a manned vehicle to be optionally manned at the commander's discretion. The applique system–equipped vehicle can be an ordinary motor vehicle in which a human retains control for manual operation or it can also support fully autonomous UGV operations.[5]

The JUAS provides many key MOSA elements (including a modular TRM, as shown in Figure 5.2), but some additional MOSA elements may be missing. For example, how are JUAS messages used at the interfaces between modules specified in JAUS TRM? Are all JUAS message sets available to all JUAS developers? In other words, are these interfaces part of an open architecture, or are some interfaces and message sets proprietary? If they are, then they may be a good candidate for the foundation of MOSA.

UMV Architecture Developments

The Navy has been employing unmanned surface and subsurface vehicles for some time in naval operations. Unmanned vehicles play a significant role in mine and counter-mine operations and undersea surveillance operations.[6] The littoral combat ship (LCS) program includes, as part of its multimission capabilities, a number of unmanned systems that can be deployed from the LCS to support a variety of missions. LCS has been designed to use modular mission packages, which can be interchanged in port in response to a change in mission, enabling this relatively small ship to support a range

---

[5]   Robotic Systems Joint Project Office, 2011a.

[6]   Craig Graham, "The Future of Maritime Warfare: Unmanned Undersea Vehicles (UUVs)," August 20, 2012..

of missions. Because of the modular nature of LCS, an open system architecture is an important element of the LCS acquisition strategy.

Another important element of the LCS open system architecture approach is the requirement for LCS-based operators to control a diverse set of unmanned systems using the same control systems (so that operator control systems do not have to be changed when mission capability packages are changed). The LCS Program Executive Office is developing a standard software architecture for unmanned system C2 to support this MOSA requirement.

JANUS Communications Standard

Navy UUV interoperability has been hampered in the past by incompatible communications systems:

> When it comes to the underwater domain . . . achieving interoperability is currently impossible due to the lack of common standards and protocols for wireless communication.[7]

According to the same source:

> Almost all underwater vehicles or sensors currently use proprietary interfaces and protocols for communication, especially for wireless communication in water.[8]

This implies that UUVs made by different manufacturers cannot communicate with each other, even if they operate in the same area, and that human operators afloat or ashore cannot control these UUVs unless they use the control systems supplied by each manufacturer.

The NATO Centre for Maritime Research and Experimentation (CMRE) is working with industry firms to develop communications standards and encouraging development of software-defined modems and networks for UUVs and UUV control systems. CMRE and industry have worked together to develop the JANUS open source communications standard for UUVs.[9] JANUS is far from a complete architecture for UUVs, but it does provide a necessary and important starting point for improving UUV interoperability.

Maritime Open Autonomy Architecture

The Navy has supported the development of several modular software architectures for USVs and UUVs that can be used to program autonomous or partially autonomous

---

[7]   Edward Lundquist, "In Search of the Standard Answer," *Proceedings Magazine,* U.S. Naval Institute, February 2014.

[8]   Lundquist, 2014.

[9]   "About Janus: JANUS Community Wiki," undated.

air, surface, and subsurface vehicles. One of these is the maritime open autonomy architecture (MOAA), which was originally developed by Draper Laboratory. In 2012, MOAA was made available to select members of industry for further development as government open source software (GOSS).[10]

MOAA is a modular and extensible autonomy framework. It includes the all-domain execution and planning technology (ADEPT).[11] This autonomy framework enables complex tasks to be broken down into lower-level, simpler tasks. The ADEPT framework is loosely associated with Colonel Boyd's observe–orient–decide–act concept (commonly referred to as the OODA loop).[12] Therefore, ADEPT should support reoccurring decisionmaking cycles that may be required for effective performance in rapidly changing, unstructured environments. The basic building block of the ADEPT autonomy framework is a planning and decisionmaking node. Just as in the OODA loop concept, situational awareness information is used to monitor the external environment and to monitor progress toward achieving mission objectives. If deviation from the preplanned mission is detected, a replanning or new plan-generation process can be triggered. Such planning can be done within a hierarchy of activities or tasks, so that a modular approach can be taken to the dynamic planning tasks for the unmanned system. MOAA uses an object-oriented design and is coded in the C++ programming language. A high-level block diagram of MOAA is shown in Figure 5.3.

Mission Oriented Operating Suite Interval Programming

The Mission Oriented Operating Suite (MOOS) is another autonomous vehicle software architecture that has been developed with support from the Navy. Its primary sponsors were the Office of Naval Research (ONR) and the National Oceanic and Atmospheric Administration, and MOOS continues to be used in ONR-related research.

The MOOS software architecture can support multiple applications, as explained below. One of these is Interval Programming (IvP) Helm, which was developed in 2004 for autonomous control on UMVs and, later, UUVs. The Naval Undersea Warfare Center has played a key role in its development.

A key design paradigm in the MOOS software architecture is the so-called backseat driver paradigm, which separates vehicle control and vehicle autonomy functions:

> The vehicle control system runs on a platform's main vehicle computer and the autonomy system runs on a separate payload computer. This separation is also referred to as the mission controller–vehicle controller interface. A primary benefit

---

[10] David Perera, "MOAA Software Goes GOSS," FierceGovernmentIT, July 16, 2012.

[11] Jason M. Furtado, "Human Interactive Mission Manager: An Autonomous Mission Manager for Human Cooperative Systems," Cambridge, Mass.: Massachusetts Institute of Technology, 2007.

[12] J. R. Boyd, "The Essence of Winning and Losing," 1995.

Figure 5.3
MOAA High-Level Framework

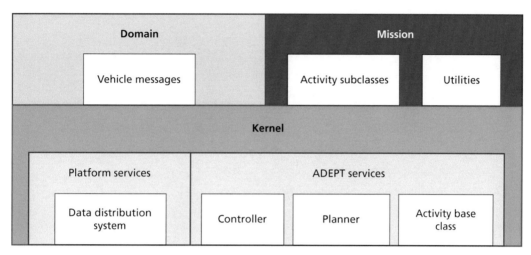

SOURCE: Furtado, 2007.
RAND *RR626-5.3*

is the decoupling of the platform autonomy system from the actual vehicle hardware. The vehicle manufacturer provides a navigation and control system capable of streaming vehicle position and trajectory information to the main vehicle computer, and accepting a stream of autonomy decisions such as heading, speed and depth in return. Exactly how the vehicle navigates and implements control is largely unspecified to the autonomy system running in the payload.[13]

This modular relationship is depicted in Figure 5.4.

MOOS follows a number of key MOSA principles in its design: modular independence; the lack of proprietary interfaces; and, modules connected using simple, well-documented interfaces. It is important to note that the MOOS software architecture is not a complete MOSA. As Figure 5.4 indicates, the vehicle navigation and control systems could be provided by the vehicle vendor and could contain proprietary software and interfaces internally. However, the autonomous aspects of the system are written completely in the MOOS architecture and depend only on nonproprietary interfaces.

Another important aspect of the MOOS software architecture is that it depends on a central publish and subscribe database. Each MOOS module communicates with

---

[13] Michael R. Benjamin, John J. Leonard, Henrik Schmidt, and Paul M. Newman, "A Tour of MOOS-IvP Autonomy Software Modules," Computer Science and Artificial Intelligence Laboratory Technical Report, MIT-CSAIL-TR-2009-006, Cambridge, Mass.: Massachusetts Institute of Technology, February 2009.

Figure 5.4
MOOS and IvP Helm Modules

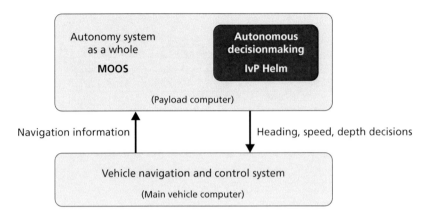

SOURCE: Benjamin et al., 2009.
RAND *RR626-5.4*

other modules using the central database, which is the core of the MOOS messaging infrastructure.

UAS Architecture Developments

Over the past decade, UAS have developed quickly. All of the military services have asked for industry support in developing and delivering a growing number of UAS with greater capabilities. Many of these new systems have been delivered as quick-reaction capabilities and not through the standard acquisition process. As a result, many of these systems do not comply with some of the detailed acquisition guidance that has been developed concurrently, including acquisition guidance for the use of modular open system architecture principles. In this section, we review some of the architectures that have been developed in industry and academia to support the insertion of autonomous capabilities into UAS. Most current UAS have relatively limited autonomous capabilities. However, academic researchers have been experimenting with and developing UAS that have greater autonomy.

Open Architecture for the Integration of UAV Civil Applications

Researchers in Europe have proposed an open architecture to integrate sensing, navigation, and control capabilities of UAVs into a single unified system. This architecture is based on an SOA approach and includes a UAV service abstraction layer (USAL). The USAL enables flight control systems and avionics to be controlled using high-level services. It abstracts the features of these systems from other components of the UAS system, including possibly an autonomous decisionmaking engine. Each node or major component of the UAV would be encapsulated by service container software and could therefore become a module of an open architecture.[14]

The set of all service containers on the UAV would compose the middleware layer of this modular architecture. This middleware layer would serve a communications bus internal to the aircraft and would also mediate communications between the aircraft and the UAV GCS. The USAL and middleware would offer a light-weight SOA with a built-in serverless communication infrastructure. The USAL would offer a large set of available services, which could be tailored to support specific UAVs and missions. The services have been classified by the authors of this UAV architecture into the four categories shown in Figure 5.5.

A potentially serious drawback of the USAL approach proposed by Pastor and his coauthors is that it appears to be closely tied to a Web services–based approach for SOAs. Although existing commercially based SOAs have good performance capabilities to support enterprise applications on enterprise networks, it is not clear that such an SOA approach would work with on a UAS platform with limited avionics bus resources. In addition, on such aircraft, many avionics systems employ embedded system processing components that have very limited input/output capabilities. At the same time, UAS flight control requirements imply a need for real-time responsiveness to sensor input and other data. Systems more tailored for real-time messaging and real-time control with embedded systems are probably preferred for UAS software archi-

Figure 5.5
UAV Service Modules

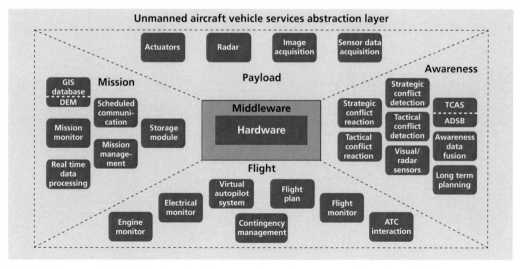

SOURCE: Pastor et al., 2009. Used under the terms of Creative Commons licensing guidelines.
RAND *RR626-5.5*

---

[14] E. Pastor, C. Barrado, P. Royo, J. Lopez, and E. Santamari, "An Open Architecture for the Integration of UAV Civil Applications," in Thanh Mung, ed., *Aerial Vehicles,* 2009.

tectures. In fact, it is probably the latter point that has slowed the development and adoption of MOSA principles within UAS programs and R&D initiatives.

Furthermore, we note that the proposed architecture does not emphasize security in communications links or in other parts of the architecture. Absent any security layer, hackers pose a threat; they could potentially spoof or jam UAS command signals, possibly enabling an adversary to take over the unmanned system. In 2009, it was discovered that insurgents successfully intercepted video feeds from unmanned platforms using cheap software to exploit the use of unencrypted data links between the unmanned system and the ground control station.[15] More recently, in April 2013 researchers successfully hacked civilian aircraft controls using smartphone technologies.[16] These examples highlight the need to incorporate security into autonomous software that can verify messages received from the system's communications links.

Data Distribution System–Based Architectures

This issue is illustrated in Figure 5.6, which shows the qualitative difference and real-time performance of different messaging and SOA-based approaches.

The object management group and open standards organization have sponsored the development of a set of open standards for the implementation of real-time enterprise service bus called the Data Distribution Service. One company in particular, Real Time Innovations Inc., has pioneered the development of DDS products. However, DDS products are available from other vendors as well. DDS is used in a number of DoD programs including unmanned system programs.

DDS is specifically designed for tactical edge applications where network bandwidth may be limited and where there may be embedded systems operating with limited Internet-like capabilities. It provides a lightweight serverless infrastructure for publish and subscribe information between dissimilar tactical edge nodes. Such nodes could, in principle, be different avionics components of a UAV for different sets of applications and communication systems connected to a UAS GCS. Shown in Figure 5.6 are software technologies that can be used to provide real-time communications for embedded systems and components with tactical edge performance requirements. Although this is a qualitative chart, its shows that Real-Time Specification for Java (RTSJ), Real-Time Common Object Request Broker Architecture (RT CORBA), and DDS messaging systems can potentially provide messaging capabilities to UAS modules that have real-time performance requirements.

---

[15] Siobhan Gorman, Yochi J. Dreazen, and August Cole, "Insurgents Hack U.S. Drones," *Wall Street Journal,* December 17, 2009; Noah Shachtman, "Insurgents Intercept Drone Video in King-Size Security Breach," *Wired,* December 17, 2009.

[16] Iain Thomson, "Researcher Hacks Aircraft Controls with Android Smartphone," *The Register,* April 13, 2013.

Figure 5.6
Message Latency for Alternative Messaging Technologies and Standards

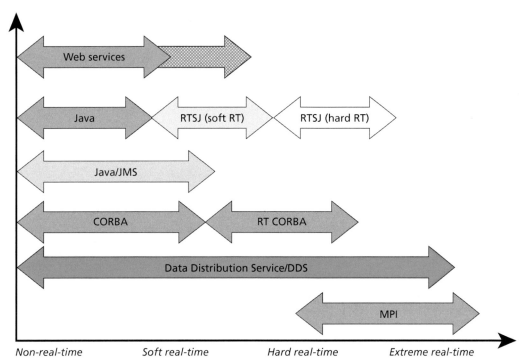

SOURCE: Gerardo Pardo-Castellote, "DDS Tutorial, 2009 OMG RT Workshop," RTI Inc., 2009. Used with permission.
RAND *RR626-5.6*

These messaging systems have performance advantages over the other more common software architecture approaches used in enterprise networks and personal computers (standard Web services and Java-based virtual machines).

DDS offers other advantages for tactical edge applications, including the ability to interoperate with components that use multiple languages and to interoperate with very simple embedded networking systems. At the same time, it can provide real-time quality of service functionality, and it also provides some of the functionality of standard Web services for publish and subscribe messaging.

As mentioned above, DDS has been adopted for use in the technical reference model for the UAS I-IPT UCS architecture. The DDS set of standards provides a service-oriented network scheme for real-time systems. Some DDS products can support network-centric data models for publish and subscribe services that mimic high-level software functions offered by relational databases and enterprise-level Web services.[17] This eliminates the need for a high-speed Web services engine at the core of the architecture.

However, DDS does not provide a complete development environment or architecture for autonomous unmanned systems. It provides only one necessary component—a real-time messaging service that can connect different avionics components, sensors, and communications payloads. In addition, DDS has been successfully deployed in a number of tactical edge systems. Therefore, it demonstrates that such a serverless SOA architecture can be implemented in such systems.

Johns Hopkins University Applied Physics Laboratory Swarming UAS Architecture

Johns Hopkins University Applied Physics Laboratory (JHU APL) has developed small, autonomous UAVs that demonstrate cooperative behaviors to accomplish simple surveillance missions. As part of this work, JHU APL developed an autonomy framework or architecture for such systems. The JHU APL UAVs could take off and land autonomously and could swarm cooperatively to detect targets. These UAVs demonstrated simple teaming arrangements between UAVs using a number of different search algorithms. Although the algorithms that were demonstrated were relatively simple, the target locations were not preprogrammed into the UAVs in advance. Mission control software was based as on a finite state automata system. The JHU APL research team also developed a multi-UAV command, control, and communications architecture that used an 802.11 wireless mobile ad hoc network. At the conclusion of this work, JHU APL proposed a UAV modular architecture that can support swarming or teaming behaviors. This architecture is illustrated in Figure 5.7.

This architecture contains two levels of control, each provided by separate modules. Low-level control processes are shown on the left, and high-level mission control processes are shown on the right. Wireless communications with other UAVs and with human pilots at GCSs are shown on the lower right and are mediated by a wireless communications layer. The JHU APL architecture was developed to support the operation of relatively small UAVs. It is also what could be considered a first-generation architecture that was developed in the early 2000s. The system depends on a single communications payload that is used for communication to the UAV GCS as well as two other UAVs. Larger, more complex UAVs now in operation carry multiple communications payloads and multiple sensors and can downlink sensor data to a GCS directly or through satellite communication links. So, as we will see in later architectures proposed by other research teams and by the USAL architecture proposed by Pastor and his coauthors, the architecture for larger UAVs must support multiple onboard sensors and communication systems. The JHU APL architecture does not make use of an SOA and probably did not need one because of the relatively simple

---

[17]   By "high-level functions" we mean software routines that use Web services description language or structured query language database calls.

Figure 5.7
JHU APL Autonomous UAV Architecture

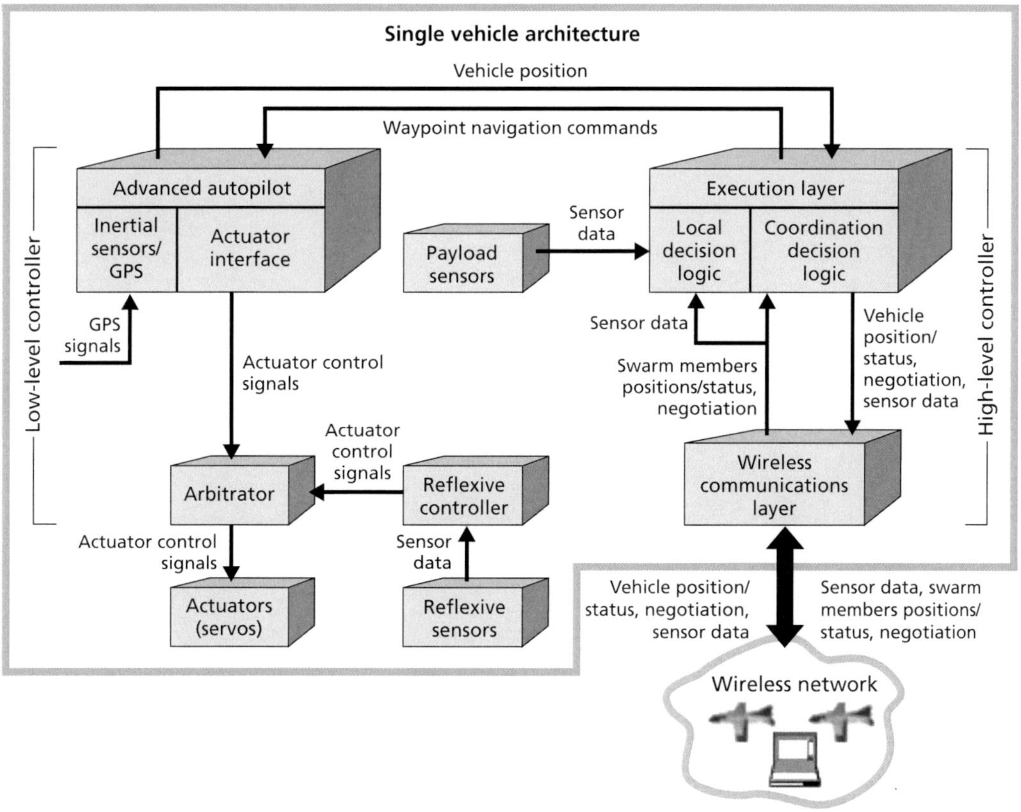

SOURCE: Robert J. Bamberger, Jr., David P. Watson, David H. Scheidt, and Kevin L. Moore, "Flight Demonstration of Unmanned Aerial Vehicle Swarming Concepts," *Johns Hopkins APL Technical Digest*, Vol. 27, No. 1, 2006. Used with permission. © The Johns Hopkins University Applied Physics Laboratory.
RAND *RR626-5.7*

and small number of payloads the JHU APL UAVs carried. More complex UAVs that carry multiple payloads will likely require one or more messaging buses and perhaps an SOA to support payload and flight control messaging.

MIT and Aurora Flight Sciences Decentralized Autonomous UAV Framework

MIT and Aurora Flight Sciences have developed and demonstrated a decentralized team of UAVs capable of cooperative search, acquisition, and track (CSAT) functions. While the primary focus of this research was to develop better algorithms to handle cooperative search and track functions, this team also developed a high-level UAV functional architecture that has some utility for thinking about how to define a modular open system architecture for UAVs in general. The CSAT architecture is shown in Figure 5.8.

Figure 5.8
Module Structure of the MIT CSAT Architecture

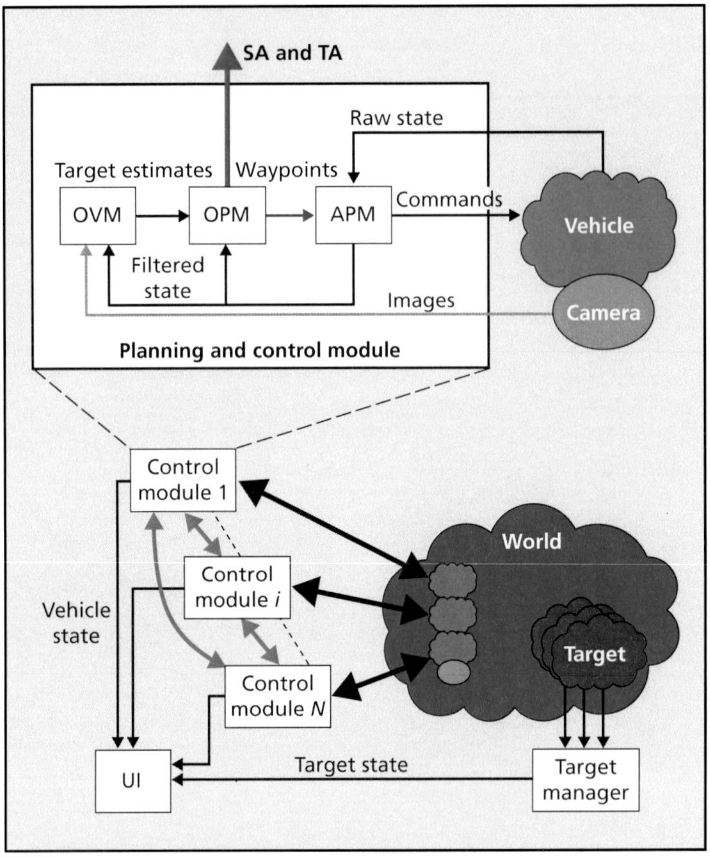

SOURCE: J. How, C. Frasher, K. C. Kulling, L. F. Bertuccelli, O. Toupet,
L. Brunet, A. Bachrach, and N. Roy, "Increasing Autonomy of UAVs,"
Robotics & Automation Magazine, IEEE 16.2, Institute of Electrical and
Electronics Engineers, 2009. Used with permission.
RAND RR626-5.8

Each UAV has a control module for planning and control of the vehicle. Because small UAVs were used in this demonstration, UAV sensor control functions were not needed. The control module had three major subelements as shown in the figure: the onboard vision module (OVM), the onboard planning module (OPM), and the auto pilot module (APM). The OPM generated situational awareness (SA) information, including target detection and target-tracking data from sensor data processed by the OVM. OPM SA data are then sent to OPMs on other UAVs and to the APM on the same vehicle, which used this information to generate navigation commands. Target estimates were sent from the OVM to the OPM, and waypoints were sent from the OPM to the APM. Vehicle state information was sent back by the APM to the other modules. Cooperative behavior was enabled by the UAVs by sharing information between UAVs (i.e., situation awareness and targeting information). The algorithm that was employed for CSAT used relatively less bandwidth than earlier cooperative tracking algorithms that were demonstrated by previous researchers. The figure also shows how moving targets were used in this demonstration. Moving targets were controlled by a target manager, which sent target state information to the UI, which could then compare the quality of the tracking information produced by the cooperative UAV fleet to the actual target locations.

MIT Lincoln Laboratory Reference Software Architecture

MIT Lincoln Laboratory (MIT/LL) has developed an architecture for data and software services for UAVs. The goal of this effort is to develop a UAV information-sharing framework for existing nonautonomous UAVs that can operate in a wide range of air domains to include the national airspace system (NAS), terminal control and landing, oceanic flight, and tactical operations.

The DoD and the Federal Aviation Administration (FAA) are undertaking a number of initiatives to safely integrate UAVs into the NAS. This means that these unmanned aircraft will have to be equipped with safety systems, such as traffic collision avoidance systems (TCAS) that are now used by human pilots on passenger aircraft. The plan is to incorporate similar sensors onto UAS so that they can provide an autonomous sense and avoid (SAA) capability.

The proposed architecture is based on SOA with open standards for key interfaces. A key feature of this architecture is the ability to accommodate SAA capabilities. An important near-term way of providing SAA is to leverage ground-based surveillance assets. The U.S. Army is leading the development of the ground-based sense and avoid (GBSAA) initiative. One longer-term solution is potentially an airborne-based sense and avoid capability in which SAA sensors located on board the UAV are used to avoid collisions with other aircraft.

Therefore, in the long term, this architecture must be able to accommodate offboard as well as onboard SAA information. This complicates the architecture because the UAV will have to interoperate with a diverse set of ground-based surveillance assets. This makes an SOA approach necessary, as it must provide the necessary mediation elements to interoperate with these ground-based systems.

Figure 5.9 shows a notional set of services that this architecture would provide. Data from sensor sources onboard and offboard the vehicle would be published to the two data buses shown in the figure. Other services such as the SAA service would subscribe to sensor data and produce their own output data products. These services would be organized into a two-layer system. Each layer would have its own quality of service requirements, with the most stringent real-time performance requirements imposed on tactical services. For this reason, the architecture uses two separate message buses to ensure quality of service for real-time and near-real-time services. MIT/LL has selected

Figure 5.9
Notional View of Possible Services in the MIT/LL Reference Architecture

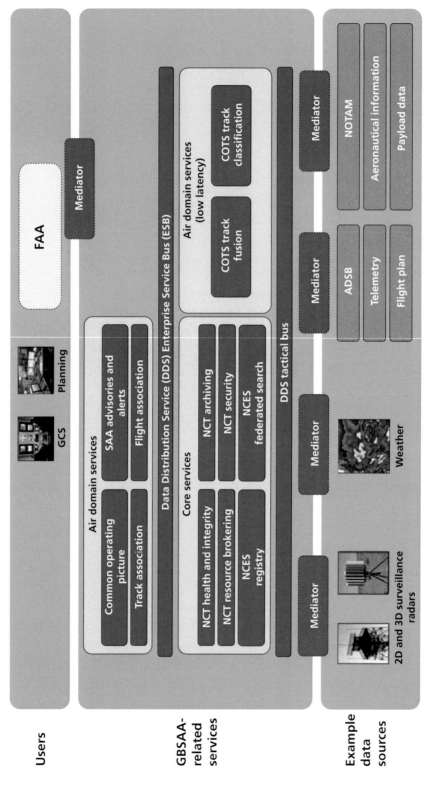

SOURCE: Curtis W. Heisey, Adam G. Hendrickson, Barbara J. Chludzinski, Rodney E. Cole, Mark Ford, Larry Herbek, Magnus Ljungberg, Zakir Magdum, D. Marquis, Alexander Mezhirov, John L. Pennell, Ted A. Roe, and Andrew J. Weinert, "A Reference Software Architecture to Support Unmanned Aircraft Integration in the National Airspace System," *Journal of Intelligent Robotic Systems*, DOI 10.1007/s10846-012-9691-8, 2012. Used under the terms of Creative Commons licensing guidelines.

DDS as the infrastructure for both the tactical data bus and the enterprise service bus. Because of the diverse nature of the sensors needed to support this architecture, a single messaging standard would not be used. In their initial development, MIT/LL has chosen to implement XML and binary format (interface definition language [IDL]) messages.

A partial list of open messaging standards in the MIT/LL reference architecture is shown in Table 5.1. One can see that this architecture relies on messaging standards defined in the UAS I–IPT UCS architecture.

Table 5.1
DoDAF TV-1 Partial List of Open Messaging Standards Used in the Reference Architecture

Service Area	Technical Services	Standards	Standard Description
Core	Command and control Resource tasking	UCI, OGS, SPS	Unmanned aerial systems C2 initiative, open geospatial consortium sensor planning service
	Algorithm advisory Track internal	New standard Army GBSAA	A proposed standard A comprehensive track schema defined in IDL to be replaced by evolving standards
	GCS internal	UCS	UCS architecture
	Track consumer Track consumer Flight plan	CoT CAT48 consumer Flight object XML	Cursor on target ASTERIX CAT48 Forthcoming standard from FOWG or EUROCONTROL
	Telemetry Weather Aeronautical information	STANAG 4586 WXXM AIXM	NATO standard Weather information exchange model Aeronautical information exchange model

SOURCE: Heisey et al., 2012.

CHAPTER SIX

Conclusions

Unmanned systems will require improved interoperability and greater autonomy to:

- operate in a cost-effective manner with fewer human personnel
- consume less communications bandwidth
- operate effectively in more demanding A2/AD environments
- operate more flexibly and to have increased survivability against increasing threats
- support a wider range of more complex missions.

However, the DSB has found that DoD acquisition programs developing unmanned systems have made limited progress in inserting greater autonomy into their products.

Study Objectives and Scope

The objectives of this study are to identify key features of unmanned systems and their associated architectures that can enable:

- improved unmanned vehicle UxV-to-UxV interoperability
- greater autonomy in unmanned systems
- cooperative UxV behaviors so that UxVs can work together in teams and accomplish complex missions in demanding A2/AD environments.

These objectives can potentially be accomplished by the adoption of open system architecture principles.

Initially, the scope of this research was set broadly to cover all types of unmanned systems: UAS, UGS, and UMS and joint or common architecture development efforts associated with each type of system. However, DoD R&D efforts and joint architecture developments were found to be less mature for UGS and UMS. In addition, the navigation problems of UGVs and UMVs differ significantly from those of UAVs. These can and should influence how architectures are defined for these systems. In addition, many important aspects of the software architectures of UGS and UMS were found to be proprietary. For these reasons, we focused our more detailed analysis efforts on UAS, and UAS architectures. We believe that further useful work can be done on UGS and UMS architectures that can improve their interoperability and autonomous capabilities in follow-on studies.

We reviewed the DoD's initiatives for improving UxV interoperability and autonomy as well as selected commercial initiatives for developing autonomous vehicles. The majority of UAS program of record architecture products are currently limited to single program views, which are not easily used to assess interoperability issues that extend across program or military service boundaries. In this study, we examine how UxV architectures can be improved so that they can support efforts to increase the degree of autonomy and interoperability of unmanned systems.

UxS Architecture Development Challenges

Findings

Our work yields several findings related to UxS architecture development:

- Current DoDAF 1.0 architecture products produced by UAS PORs do not contain the information needed for joint interoperability analysis or the information needed to predict the occurrence of joint interoperability problems.
- There are many significant differences in the designs, capabilities, and functions of UxS that operate in different domains (air, land, and sea). These differences imply that the DoD should not try to develop a single common UxS architecture for all types of unmanned systems.
- UMS and UGS developments are not as mature as UAS developments. The DoD has developed a larger number of UAS with a wide range of operational capabilities than UGS or UMS.
- The UAS Task Force I-IPT Horizontal Integration Working Group (HIWG) has developed an initial version of an as-is JCUA that has proven useful in identifying and documenting UAS gaps.[1]
- Another UAS I-IPT working group has developed the UCS architecture, which provides a common UAS GCS architecture designed to improve UAS-GCS cross program interoperability.
- UCS and JCUA architectures may not be aligned.
- The DoD does not have a common syntax for UAS, UGS, or UMS architectures.

---

[1]   Frawley, 2012.

Recommendations

Given these findings, we offer the DoD several recommendations. First, we recommend forgoing developing a joint common UAS architecture. Instead, we recommend that the DoD pursue the federation of existing UAS acquisition program architectures, including those developed, maintained, and used by individual UAS programs of record.

Can such architectures be federated (or made compatible), so they can predict interoperability problems, without having a "central joint target" that individual service programs can use to guide system developments? We believe that a central joint target is needed and recommend that this be based on the UCS architecture. Thus, it should be one of the architectures that is federated because it relies on open standards, has a well-defined modular structure and TRM, and because we found that it has already been useful in demonstrating cross-program UAS interoperability.

However, because of the many differences between UxS that operate in different domains, we do not recommend that either a common or federated architecture be developed for all UxS.

A Common UAS Architecture Syntax

Findings

The DoD does not have a common syntax for UAS, UGS, or UMS architectures.

Recommendations

We recommend that the DoD incrementally develop common joint semantics for key architecture data elements. Because of the complexity of this task, it should be approached in a top-down manner where the highest-priority modules or functional elements of the architecture are sequentially addressed in efforts to develop a common syntax. The key elements in a common taxonomy should define (at least at a top level) UAS capabilities, components or subsystems, commands or messages, and mission elements.

If this is done, these could then be used to express joint mission threads (JMTs) associated with these architectures and also to incorporate service POR architecture products into the Joint Architecture Federation and Integration Project (JAFIP). This is important because JAFIP will work only if a common syntax and data standards are developed; otherwise, a federation is not possible. In the short term, efforts should focus on semantics to make the sharing proposed by JAFIP work properly.

Given the difficulty in creating a comprehensive common syntax for all UAS and the time it would take to reach agreement among all programs and players, a common syntax should be developed incrementally that first covers major UAS architecture data elements. In this way the number of translated terms can be minimized eventually over time.

Additional Steps to Enable UxV Architecture Federation and Autonomy

Findings
Our review of high-level unmanned system architectures and frameworks that have been used to in prior research efforts reveals that these frameworks are dissimilar for unmanned systems that operate in different domains (air, maritime, and ground).

Recommendations
We recommend that a POSA framework for a joint integrated UAS architecture be developed to support the federation of UAS architectures. This will not only improve unmanned system interoperability and component reuse but could also improve system autonomy. If this framework is designed properly, it could also be used to improve the autonomy of unmanned systems and enable new autonomous capabilities to be readily inserted into such systems (if open interfaces are used).

We recommend that UxS architecture federation efforts be pursued independently in each domain (land, sea, and air). The mission context for each class of systems and the environmental factors that each class of system must contend with are so different that they may lead to dissimilar architectures and frameworks.

UAS Modular Technical Reference Model

Our review of software architectures for autonomous or semiautonomous unmanned systems reveals that different development teams have selected different modular schemes for their architectures. In addition, they have chosen different software foundations for their software architectures. This extends to the messaging standards and messaging approach that are used. For example, some software architectures rely on a centralized database or server for the messaging infrastructure, but others do not. Architecture frameworks that appear to be used most frequently in UAVs fall into the latter category. The advantage of a decentralized communications bus is that they generally have real-time messaging performance, which is important for maintaining real-time control of a fast-moving vehicle. An example of the decentralized approach is the OMG DDS, which is used in many real-time tactical systems in the DoD.

Some additional important architectural constructs have been proposed in academic research projects that have developed autonomous UAVs. One such concept is that of multiple levels of control, an approach that has been used by a number of different research teams (JHU APL UAV architecture, the MIT CSAT architecture, and the more recent UAV architecture proposed by MIT Lincoln Laboratory).

In Figure 6.1, we identify what we believe are the essential modular components of a UAV TRM that can support multiple levels of autonomy and that is consistent with the UAS POSA approach described above. This proposed TRM includes features that have been identified in the UAV high-level architectures developed and, in many cases, demonstrated by different research teams. The proposed TRM supports two levels of system control. Highly responsive vehicle control would be accomplished using the APM, which would be connected to flight control systems and sensors using a high-speed tactical data bus. This data bus would deliver messages with a high assurance of low or minimal time delay to enable real-time control and feedback loops. Other services or modules that require real-time performance would also be connected to the same tactical data bus. We recommend that the tactical service bus use the DDS standard, as indicated in the figure, to ensure interoperability with the UCS architecture.

The system would be equipped with the second enterprise service bus to support services that did not have real-time communications requirements. Examples of these are shown in the figure, including air domain services that would process air track information and SAA advisories from offboard C2 or air surveillance centers. In addition, mission sensors that produced a high volume of mission data also may be connected using the enterprise service bus.

Figure 6.1
Proposed UAS Modular Technical Reference Model (TRM)

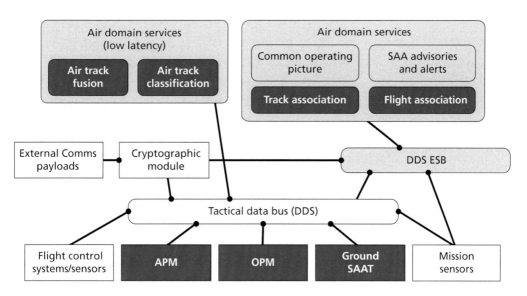

The modules in the TRM that could be enabled with autonomous capability are shown in blue in the figure. These modules would be connected to the overall system using open interfaces. This would enable these modules to be produced by outside contractors that have special expertise in autonomous systems. The TRM would enable autonomy to be inserted into four key areas in the UAS architecture: (1) the real-time APM, (2) the ground situational awareness and targeting module (SAAT), (3) the OPM, and (4) the low latency air track management services.

With further development and specification of the open interfaces that connect these modules to the larger system, this TRM will enable UAS programs to comply with the design recommendation made by the DSB, namely, to separate the autonomous capabilities of unmanned systems from the rest of the vehicle platform.

All of the modules highlighted in blue could reduce UAV communications demands significantly and could eliminate the need for real-time communications to the vehicle for remote pilot control in A2/AD environments. For example, air track would not have to be sent up to or down from the aircraft to identify potential aircraft or terrain collision events.

It should also be noted that the TRM shown in the figure includes all of the modules needed by a large, high-value UAV that could fly in U.S. FAA-controlled airspace and which may be equipped with defensive countermeasures against aircraft threats. Smaller UAVs that fly at low altitude, that would not need to fly in FAA controlled airspace, and that would be too small to be equipped with aircraft sensors and warning systems would not need the modules highlighted in purple in the figure.

UGV and UMV MOSA Frameworks

Should POSAs and TRMs for UGVs and UMVs be similar in composition and scope to the one proposed for UAS? We believe that the answer is no. Unmanned systems are robots that are increasingly going to be programmed to operate semiautonomously in specific ways with warfighters and threats in distinct environments, just as ships, ground vehicles, and even traditional manned aircraft interact in fundamentally different ways.[2] We found evidence of these distinctions when examining the different architectures and modular designs of unmanned systems designed to operate in these different domains. Our review of high-level unmanned system architectures and frameworks reveals that these frameworks are dissimilar for unmanned systems that operate in different domains (air, maritime, and ground). The mission context, control systems for each class of system, and the environmental factors that each class of system must contend with are so different that they may lead to dissimilar architectures and

---

[2] We recognize that manned aircraft have an autopilot capability, which is a form of semiautonomy. However, in this case, the pilot is always in the aircraft and can assume control. In an unmanned system, pilots operate the system remotely and, furthermore, some UAS are launch-on-mission with no interaction until the UAS returns.

frameworks. Therefore, we recommend that POSA development efforts be independently pursued for UxS that operate in different domains. We believe that a tailored POSA should be developed for UGS, and a separate UMS POSA should be developed for USVs and UUVs.

Proposed Next Steps

Even though it may be premature to develop common or federated architectures for UGS and UMS, developments of common syntaxes for UGS and UMS architectures should begin soon. This will make future joint architecture developments easier regardless of which approach is eventually chosen for UGS and UMS.

Further research should be conducted on autonomous UGVs and UMVs and on the architectural constructs used by developers of these systems before TRMs are developed for these classes of unmanned systems. Several important DARPA and Navy programs are nearing development of initial conceptual designs for USVs and UMVs. Information from these programs can be used to develop a TRM for a joint common unmanned maritime system architecture that is based on the latest software and robot technologies.

In addition, Google and commercial automobile manufacturers are developing proprietary autonomous vehicle systems and technologies. The underlying architectures for these systems are proprietary and have not been made available to the open source software development community. However, it may be possible, with assistance from major industry firms, to develop a UGV POSA that can be used to advance military UGV development, interoperability, and autonomy that still preserves the intellectual property and R&D investments of private firms. Research in this area could be conducted by a federally funded research and development center if appropriate nondisclosure agreements are negotiated with these private firms.

In our analysis, we do not investigate in detail the specific software architectures used by the different development teams that have explored or developed autonomous unmanned systems. Important insights can be gained by an examination of the details of the software architectures and by comparing them. For example, should the TRM include specified software development environments and SDKs? A related question is whether open source software code bases should be used in specific parts of an open modular architecture for UAS.

And, finally, the open interfaces should be defined for the key modules in the proposed UAS TRM that are designed to contain autonomous capabilities. These open interfaces could be established by examining the messaging formats and communication buses used by leading research teams and by interviewing autonomy experts in the DoD R&D community and industry.

References

"About Janus: JANUS Community Wiki," undated. As of February 17, 2014:
http://www.januswiki.org/tiki-index.php?page=About+Janus

"ACTUV Program Initiates Concept Designs," DARPA press release, December 2010.

Anderson, James M., Nidhi Kalra, Karlyn, D. Stanley, Paul Sorensen, Constantine Samaras, and Oluwatobi A. Oluwatola, *Autonomous Vehicle Technology: A Guide for Policymakers*, Santa Monica, Calif.: RAND Corporation, RR-443-1-RC, 2014. As of April 21, 2014:
http://www.rand.org/pubs/research_reports/RR443-1.html

Architect Role and Developer Role, "The DoDAF Architecture Framework Version 2.0," 2011. As of June 9, 2014:
http://citeseerx.ist.psu.edu/viewdoc/download?doi=10.1.1.190.7841&rep=rep1&type=pdf

Ashton, Captain Duane, "Unmanned Maritime Systems Autonomy," presentation delivered at the 10th International MIW Technology Symposium, PMS 406 Unmanned Maritime Systems, May 2012.

Axe, David, "Did Iran Capture a U.S. Stealth Drone Intact?" *Wired* (online), December 4, 2011. As of April 21, 2014:
http://www.wired.com/dangerroom/2011/12/did-iran-capture-a-u-s-stealth-drone-intact/

Bamberger, Robert J., Jr., David P. Watson, David H. Scheidt, and Kevin L. Moore, "Flight Demonstrations of Unmanned Aerial Vehicle Swarming Concepts," *Johns Hopkins APL Technical Digest*, Vol. 27, No. 1, 2006.

Barber, Daniel, "Joint Architecture for Unmanned Systems," University of Central Florida, Institute for Simulation and Training, January 29, 2011.

Benjamin, Michael R., John J. Leonard, Henrik Schmidt, and Paul M. Newman, "A Tour of MOOS-IvP Autonomy Software Modules," Computer Science and Artificial Intelligence Laboratory Technical Report, MIT-CSAIL-TR-2009-006, Cambridge, Mass.: Massachusetts Institute of Technology, February 2009.

Boyd, J., "A Discourse on Winning and Losing," Maxwell AFB, Ala: Air University Library, Document No. M-U 43947, August 1987.

———, "The Essence of Winning and Losing," 1995. As of April 21, 2014:
http://www.danford.net/boyd/essence.htm

Button, Robert W., John Kamp, Thomas B. Curtin, and J. A. Dryden, *A Survey of Missions for Unmanned Undersea Vehicles,* Santa Monica, Calif.: RAND Corporation, MG-808-NAVY, 2009. As of April 16, 2014:
http://www.rand.org/pubs/monographs/MG808.html

Chief Information Officer, U.S. Department of Defense, "The DoDAF Architecture Framework Version 2.02," August 2010. As of September 11, 2013:
http://dodcio.defense.gov/dodaf20.aspx

———, "The DoDAF Architecture Framework Version 2.02," January 27, 2011. As of April 14, 2014:
http://citeseerx.ist.psu.edu/viewdoc/download?doi=10.1.1.190.7841&rep=rep1&type=pdf

Cordova, Amado, Lindsay Millard, Lance Menthe, Robert A Guffey, and Carl Rhodes, *Motion Imagery Processing and Exploitation (MIPE)*, Santa Monica, Calif.: RAND Corporation, RR-154-AF, 2013. As of April 16, 2014:
http://www.rand.org/pubs/research_reports/RR154.html

DARPA, "Grand Challenge Overview," March 2004. As of April 21, 2014:
http://archive.darpa.mil/grandchallenge04/overview.htm.

"DARPA Grand Challenge," *Wikipedia*, undated. As of January 30, 2014:
http://en.wikipedia.org/wiki/DARPA_Grand_Challenge

Defense Science Board, "Task Force Report: The Role of Autonomy in DoD Systems," Washington, D.C.: Office of the Secretary of Defense for Acquisition, Technology, and Logistics, July 2012. As of April 14, 2014:
http://www.acq.osd.mil/dsb/reports/AutonomyReport.pdf

Department of Defense, Procedures for Interoperability and Supportability of Information Technology (IT) and National Security Systems (NSS), June 30, 2004. As of April 14, 2014:
http://jitc.fhu.disa.mil/jitc_dri/pdfs/i46308.pdf

———, "Department of Defense Global Information Grid Architecture Federation Strategy," Version 1.2, August 7, 2007.

———, Interoperability–Integrated Product Team, "Mission Statement," 2010. As of April 15, 2014:
http://www.interoperabilityipt.org/page/mission

———, "Unmanned Systems Integrated Roadmap FY2011–2036," 2011. As of August 29, 2013:
http://www.acq.osd.mil/sts/docs/Unmanned%20Systems%20Integrated%20Roadmap%20FY2011-2036.pdf

———, "DoD Open Systems Architecture (OSA) Contract Guidebook for Program Managers," June 2013a.

———, "Unmanned Systems Integrated Roadmap FY2013–2038," December 2013b. As of June 9, 2014:
http://www.defense.gov/pubs/DOD-USRM-2013.pdf

Department of Defense Instruction (DoDI) 4630.8, *Procedures for Interoperability and Supportability of Information Technology (IT) and National Security Systems (NSS)*, enclosure 2, Washington, D.C.: Department of Defense, June 30, 2004. As of April 14, 2014:
http://jitc.fhu.disa.mil/jitc_dri/pdfs/i46308.pdf

"Esri Geoportal Server | Open Source Metadata Management," undated. As of February 16, 2014:
http://www.esri.com/software/arcgis/geoportal

Federal Agencies Ad Hoc Autonomy Levels for Unmanned Systems Working Group Participants, "Autonomy Levels for Unmanned Systems (ALFUS) Framework, Volume I: Terminology, Version 1.1," NIST Special Publication 1011, September 2004, p. 14.

Frawley, Chuck, "Horizontal Integration Working Group (HIWG) Update," presentation to the Interoperability Integrated Product Team, July 18, 2012.

Furtado, Jason M., "Human Interactive Mission Manager: An Autonomous Mission Manager for Human Cooperative Systems," Cambridge, Mass.: Massachusetts Institute of Technology, 2007.

Google Developers, "KML Tutorial–Keyhole Markup Language," November 14, 2013. As of January 29, 2014:
https://developers.google.com/kml/documentation/kml_tut

Gorman, Siobhan, Yochi J. Dreazen, and August Cole, "Insurgents Hack U.S. Drones," *Wall Street Journal,* December 17, 2009.

Graham, Craig, "The Future of Maritime Warfare: Unmanned Undersea Vehicles (UUVs)," August 20, 2012. As of April 14, 2014:
http://mscconference.wordpress.com/2012/08/20/the-future-of-maritime-warfare-unmanned-undersea-vehicles-uuvs-2/

Grant, Rebecca. "Refueling the RPAs," *Air Force Magazine*, March 2012. As of April 14, 2014:
http://www.airforcemag.com/MagazineArchive/Documents/2012/March%202012/0312RPA.pdf

Green, CDR Pat, UAS Interoperability IPT, presentation, July 17, 2013.

Guizzo, Erico, "How Google's Self-Driving Car Works," *IEEE Spectrum*, October 2011. As of April 14, 2014:
http://spectrum.ieee.org/automaton/robotics/artificial-intelligence/how-google-self-driving-car-works

Heisey, Curtis W., Adam G. Hendrickson, Barbara J. Chludzinski, Rodney E. Cole, Mark Ford, Larry Herbek, Magnus Ljungberg, Zakir Magdum, D. Marquis, Alexander Mezhirov, John L. Pennell, Ted A. Roe, and Andrew J. Weinert, "A Reference Software Architecture to Support Unmanned Aircraft Integration in the National Airspace System," *Journal of Intelligent Robotic Systems,* DOI 10.1007/s10846-012-9691-8, 2012.

How, J. P., C. Fraser, K. C. Kulling, L. F. Bertuccelli, O. Toupet, L. Brunet, A. Bachrach, and N. Roy, "Increasing Autonomy of UAVs," *Robotics & Automation Magazine,* IEEE 16.2, Institute of Electrical and Electronics Engineers, 2009, pp. 43–51.

I-IPT Interoperability–Integrated Product Team, Mission Statement. As of September 11, 2013:
http://www.interoperabilityipt.org/page/mission

Kingston, Derek, "UAV Autonomy," September 19, 2012. As of September 11, 2013:
http://arclab.engin.umich.edu/wp-content/uploads/2012/09/kingston_max_annual_review2012.pdf

"K-MAX Crashes on a Mission in Afghanistan," Defense Update, June 17, 2013. As of January 30, 2014:
http://defense-update.com/20130617_k-max-crashes-on-a-mission-in-afghanistan.html

Lavrinc, Damon. "Nissan Promises to Deliver Autonomous Car by 2020," *Wired.com*, August 27, 2013. As of January 31, 2014:
http://www.wired.com/autopia/2013/08/nissan-autonomous-drive/

Lockheed-Martin, "K-MAX–Lockheed Martin," undated. As of January 30, 2014:
http://www.lockheedmartin.com/us/products/kmax.html

Lundquist, Edward, "Open Business Model," *Seapower Magazine,* November 2013.

———, "In Search of the Standard Answer," *Proceedings Magazine*, U.S. Naval Institute, February 2014.

MicroPilot, "MicroPilot–Products–MP2128g," 2011. As of January 30, 2014:
http://www.micropilot.com/products-mp2128g.htm

ODASD-SE, "DoD Systems Engineering–Initiatives," December 2013. As of April 14, 2014:
http://www.acq.osd.mil/se/initiatives/init_osa.html

Pardo-Castellote, Gerardo, "DDS Tutorial, 2009 OMG RT Workshop," RTI Inc., 2009.

Pastor, E., C. Barrado, P. Royo, J. Lopez, and E. Santamari, "An Open Architecture for the Integration of UAV Civil Applications," in Thanh Mung, ed., *Aerial Vehicles*, 2009. As of April 14, 2014:
http://www.intechopen.com/books/aerial_vehicles/an_open_architecture_for_the_integration_of_uav_civil_applications

Perera, David, "MOAA Software Goes GOSS," FierceGovernmentIT, July 16, 2012. As of September 11, 2013:
http://www.fiercegovernmentit.com/story/moaa-software-goes-goss/2012-07-16#ixzz2dbnVvbq5

Piazza, William, "Joint Architecture Federation and Integration Project (JAFIP)," presentation to the Interoperability Integrated Product Team, July 18, 2012.

Raytheon Company, "Miniature Air Launched Decoy (MALD)," undated. As of June 6, 2014:
http://www.raytheon.com/capabilities/products/mald/

Robotic Systems Joint Project Office, "Unmanned Ground Systems Roadmap," July 2011a. As of September 11, 2013:
http://www.rsjpo.army.mil/images/UGS_Roadmap_Jul11_r1.pdf

———, "UGV Interoperability Profile (IOP)–Overarching Profile JAUS Profiling Rules," Version 0, SFAE-GCS-UGV MS 266, December 21, 2011b.

———, "Unmanned Ground Vehicle (UGV) Interoperability Profile (IOP) Overarching Profile," Version 0, SFAE-GCS-UGV MS 266, December 21, 2011c.

Rockwell Collins, "Athena 311 Integrated Flight Control System," 2014. As of January 30, 2014:
http://www.rockwellcollins.com/sitecore/content/Data/Products/Controls/Flight_Controls/Athena_311_Integrated_Flight_Control_System.aspx

Rowe, Steve, and Christopher R. Wagner, "An Introduction to the Joint Architecture for Unmanned Systems (JAUS)," Cybernet Systems Corporation, November 2008.

Savitz, Scott, Irv Blickstein, Peter Buryk, Robert W. Button, Paul DeLuca, James Dryden, Jason Mastbaum, Jan Osburg, Philip Padilla, Amy Potter, Carter C. Price, Lloyd Thrall, Susan K. Woodward, Roland J. Yardley, and John M. Yurchak, *U.S. Navy Employment Options for Unmanned Surface Vehicles (USVs)*, Santa Monica, Calif.: RAND Corporation, RR-384-NAVY, 2013. As of April 22, 2014:
http://www.rand.org/pubs/research_reports/RR384.html

Shachtman, Noah, "Insurgents Intercept Drone Video in King-Size Security Breach," *Wired*, December 17, 2009.

Szoldra, Paul, "Drone Spying Capabilities Are About to Take Another Huge Leap," *Business Insider,* January 29, 2013. As of April 14, 2014:
http://www.businessinsider.com/darpa-argus-mega-camera-most-detailed-surveillance-camera-in-world-2013-1?op=1

"The DoDAF Architecture Framework Version 2.0," 2011. As of June 9, 2014:
http://dodcio.defense.gov/Portals/0/Documents/DODAF/DoDAF_v2-02_web.pdf

Thomson, Iain, "Researcher Hacks Aircraft Controls with Android Smartphone," *The Register*, April 13, 2013.

Wade, Robert, J., "Joint Architecture for Unmanned Systems," Research, Development & Engineering Command, Aviation and Missile Research, Development and Engineering Center (AMRDEC), Software Engineering Directorate, 2006.

Warwick, Graham. "Unmanned K-Max Gets Cleverer," *Aviation Week,* August 11, 2013. As of April 14, 2014:
http://www.aviationweek.com/Blogs.aspx?plckPostId=Blog:27ec4a53-dcc8-42d0-bd3a-01329aef79a7Post:31459159-9eba-46b3-9c6b-fc3de70ab6cc

Williams, Paul, and Michael Crump, "Intelligent Landing System for Landing UAVs at Unsurveyed Airfields," *Proceedings of the 28th International Congress of the Aeronautical Sciences*, 2012. As of April 14, 2014:
http://www.icas-proceedings.net/ICAS2012/PAPERS/131.PDF

Zachman, John A., "A Framework for Information Systems Architecture," *IBM Systems Journal*, Vol. 26, No. 3, 1987, pp. 276–292.